计算机系列教材

邱晓红 吴沧海 主 编
刘秋明 杨 俊 副主编

计算机专业英语
（第2版）

清华大学出版社
北京

内 容 简 介

"计算机专业英语"是计算机及相关专业的一门专业基础课程。由于计算机相关核心技术大部分源于英语国家，而且其更新的速度越来越快，如果不掌握计算机专业英语，势必受到语言的制约，严重影响对新技术的理解和消化。本书基于 CDIO 的工程教育理念，结合需要掌握的专业知识点，选取了计算机基础、计算机网络、云计算、数据库、程序设计语言、MSDN、编译原理、离散数学、软件工程、嵌入式系统、数据结构与算法、操作系统等相关内容，增强了计算机相关领域的知识点和英汉专业术语的对应关系。素材取自国内外最近几年计算机科学各个领域的最新教材、专著、论文和网络信息。本书每个单元的内容相对独立，包括课文 A、课文 B、单词、难句分析、语法、技能训练和阅读材料，对较偏、较难的单词进行了注释。修订版增加了人工智能、科技论文写作、参考译文等内容。

本书可以作为教学研究型高等院校和教学(应用)型大学计算机相关专业的专业英语教材，也可供计算机专业人员及其他有兴趣的读者学习参考。

本书封面贴有清华大学出版社防伪标签，无标签者不得销售。
版权所有，侵权必究。举报：010-62782989，beiqinquan@tup.tsinghua.edu.cn。

图书在版编目(CIP)数据

计算机专业英语/邱晓红，吴沧海主编. —2 版. —北京：清华大学出版社，2019（2024.7重印）
（计算机系列教材）
ISBN 978-7-302-52399-4

Ⅰ. ①计… Ⅱ. ①邱… ②吴… Ⅲ. ①电子计算机－英语－教材 Ⅳ. ①TP3

中国版本图书馆 CIP 数据核字(2019)第 041433 号

责任编辑：白立军　常建丽
封面设计：常雪影
责任校对：焦丽丽
责任印制：丛怀宇

出版发行：清华大学出版社
网　　址：https://www.tup.com.cn，https://www.wqxuetang.com
地　　址：北京清华大学学研大厦 A 座　　　　邮　编：100084
社 总 机：010-83470000　　　　　　　　　　邮　购：010-62786544
投稿与读者服务：010-62776969，c-service@tup.tsinghua.edu.cn
质量反馈：010-62772015，zhiliang@tup.tsinghua.edu.cn
课件下载：https://www.tup.com.cn，010-83470236

印 装 者：三河市龙大印装有限公司
经　　销：全国新华书店
开　　本：185mm×260mm　　印　张：17.25　　字　数：398 千字
版　　次：2011 年 9 月第 1 版　2019 年 8 月第 2 版　　印　次：2024 年 7 月第 7 次印刷
定　　价：45.00 元

产品编号：080914-01

前　言

计算机相关核心技术大部分源于英语国家,软件开发工具的主流及程序语言基本要素是英语。信息技术日新月异地发展,如果不掌握计算机专业英语,势必受到语言的制约,严重影响对新技术的理解和消化。所以,从事 IT 行业的人员都需要掌握计算机英语的常用术语和缩略语、常见的语法和惯用法,能借助字典熟练阅读英文文档和技术资料。"计算机专业英语"是计算机及相关专业的一门专业基础课程。

学习计算机专业英语,最好的教学方法是对计算机相关专业课采用双语教学,但教学研究型和教学(应用)型大学的学生,英语的听、说、写能力相对偏弱,双语教学对专业知识的掌握效果不那么理想。所以,为了便于学生掌握计算机专业英语,只能退而求其次,通过选择有趣、较易懂的专业素材或软件工具相关实用内容,让学生重温专业课程的知识点,并学会相关的中英文专业词汇。专业英语教学课程成为各门专业课程知识点融会的平台和载体,体现了 CDIO(构思(Conceive)、设计(Design)、实现(Implement)和运作(Operate))的工程教学理念,摆脱了双语教学难以掌握专业知识的尴尬,也增强了学习专业英语的趣味和实用性,进一步增强了计算机相关领域的应用知识点和英汉专业术语的对应关系。

本书针对教学研究型和教学(应用)型大学的学生英语素质特点,结合需要掌握的专业知识点,从国内外最近几年计算机相关领域的最新教材、专著、论文和网络信息中选取了计算机基础、计算机网络、云计算、数据库、程序设计语言、MSDN、编译原理、离散数学、软件工程、嵌入式系统、数据结构与算法、操作系统等内容素材。全书共分为 11 个单元,每个单元的内容相对独立,包括课文 A、课文 B、单词、难句分析、语法、技能训练和阅读材料,对较偏、难、专业的词汇进行了注释。修订版增加了人工智能、科技论文写作、参考译文等内容。授课教师可根据课时要求及专业特点加以裁减。

邱晓红构想了全书内容和要求,审查各单元内容并统一了全书风格,吴沧海编写了第 4~10 单元及附录 F,刘秋明编写了第 1~3 单元、第 11 单元及附录 G,杨俊编写了附录相关内容,赖谷丹、杨欣参与了本书的素材收集和整理工作。全书由邱晓红统稿并定稿。

本书内容的工程教学理念构想来自"基于布鲁姆教学理论的计算机专业外语课程教学模式的研究和探讨""教学与科研有机结合提高教学质量和学生综合素质的研究(2008B2ZB04)""实施本科生全程导师制教育培养模式的研究——以江西省高校软件学院为例(08YB078)""软件工程融合传统专业复合型创新人才培养模式研究(JXJG-12-6-

15)"等教改课题的研究成果,在此特别感谢江西省教育厅、江西理工大学等有关单位领导和教学管理部门的支持。在本书的编写过程中,我们还参阅了大量的专业英语书籍、英语网站和有关教学研究论文资料,在此向有关作者表示衷心的感谢!在本书的使用过程中,得到许多老师和同学的指正,清华大学出版社对本书的修订出版工作给予了大力的支持,在此表示衷心的感谢!

书中难免有不妥和错误之处,恳请广大读者及同行专家批评指正,以便于再版修订完善。

编 者

2018年12月于江西南昌

目　　录

Unit 1　Computer Structure ·· 1
　Text A　The Function of Computer ·· 1
　　　　　Words ·· 9
　　　　　Phrases ·· 11
　　　　　Exercises ·· 12
　　　　　批注 ·· 13
　Text B　The Future of Computer Technology ··· 14
　　　　　Words ·· 18
　　　　　Phrases ·· 19
　　　　　批注 ·· 19
　　　　　Associated Reading ·· 20
　　　　　批注 ·· 20

Unit 2　Programming Language ··· 22
　Text A　Functions in C ··· 22
　　　　　Words ·· 24
　　　　　Phrases ·· 24
　　　　　Exercises ·· 24
　　　　　批注 ·· 25
　Text B　J2ME Related ··· 26
　　　　　Words ·· 29
　　　　　Phrases ·· 29
　　　　　批注 ·· 30
　　　　　Associated Reading ·· 30

Unit 3　Discrete Mathematics ··· 34
　Text A　About Discrete Mathematics ··· 34
　　　　　Words ·· 37
　　　　　Phrases ·· 38
　　　　　Exercises ·· 39
　　　　　批注 ·· 40
　Text B　Tree ·· 40
　　　　　Words ·· 44

            Phrases ·· 44

            批注 ·· 44

            Associated Reading ·· 45

Unit 4　Software Engineering ·· 47

    Text A　Software Processes ·· 47

            Words ·· 51

            Phrases ·· 52

            Exercises ·· 53

            批注 ·· 54

    Text B　Introducing the UML ··· 54

            Words ·· 58

            Phrases ·· 59

            批注 ·· 60

            Associated Reading ·· 60

Unit 5　Database ·· 63

    Text A　MySQL Introduction ··· 63

            Words ·· 65

            Phrases ·· 65

            Exercises ·· 66

            批注 ·· 67

    Text B　Performing Transactions ··· 67

            Words ·· 69

            Phrases ·· 70

            批注 ·· 70

            Associated Reading ·· 70

Unit 6　Embedded System ··· 72

    Text A　What is an Embedded System? ··· 72

            Words ·· 75

            Phrases ·· 76

            Exercises ·· 77

            批注 ·· 78

    Text B　Getting to Know the Hardware ··· 79

            Words ·· 82

            Phrases ·· 83

            批注 ·· 83

	Associated Reading	………………………………………	84

Unit 7　Computer Network ……………………………………………… 86

- Text A　Internet Protocol Suite …………………………………… 86
  - Words ………………………………………………………… 89
  - Phrases ……………………………………………………… 90
  - Exercises …………………………………………………… 91
  - 批注 …………………………………………………………… 92
- Text B　Cloud Computing ………………………………………… 93
  - Words ………………………………………………………… 96
  - Phrases ……………………………………………………… 97
  - 批注 …………………………………………………………… 97
  - Associated Reading ……………………………………… 98

Unit 8　Data Structure …………………………………………………… 100

- Text A　Data Structures and Algorithms in Java ……………… 100
  - Words ………………………………………………………… 102
  - Phrases ……………………………………………………… 103
  - Exercises …………………………………………………… 103
  - 批注 …………………………………………………………… 104
- Text B　The Introduction of Two Important Data Structures ……… 104
  - Words ………………………………………………………… 106
  - Phrases ……………………………………………………… 107
  - 批注 …………………………………………………………… 107
  - Associated Reading ……………………………………… 107
  - 批注 …………………………………………………………… 109

Unit 9　Microsoft Developer Network ………………………………… 110

- Text A　What is MSDN? …………………………………………… 110
  - Words ………………………………………………………… 112
  - Phrases ……………………………………………………… 113
  - Exercises …………………………………………………… 114
  - 批注 …………………………………………………………… 115
- Text B　Getting to Know MSDN …………………………………… 115
  - Words ………………………………………………………… 118
  - Phrases ……………………………………………………… 118
  - 批注 …………………………………………………………… 119
  - Associated Reading ……………………………………… 119

· V ·

Unit 10　Compilers Principles ················· 121
　　Text A　The Science of Code Optimization ················· 121
　　　　　　Words ················· 123
　　　　　　Phrases ················· 124
　　　　　　Exercises ················· 124
　　　　　　批注 ················· 125
　　Text B　Optimizations for Computer Architectures ················· 125
　　　　　　Words ················· 128
　　　　　　Phrases ················· 128
　　　　　　批注 ················· 128
　　　　　　Associated Reading ················· 129

Unit 11　Operating System ················· 136
　　Text A　Operating System Overview ················· 136
　　　　　　Words ················· 139
　　　　　　Phrases ················· 140
　　　　　　批注 ················· 140
　　　　　　Exercises ················· 140
　　Text B　BIOS or CMOS Setup ················· 141
　　　　　　批注 ················· 143
　　　　　　Words and Phrases ················· 143
　　　　　　Associated Reading ················· 144
　　　　　　批注 ················· 148

附录 A　计算机专业英语主要句型及翻译技巧 ················· 149

附录 B　计算机专业英语的特点 ················· 161

附录 C　常见计算机英语缩写 ················· 172

附录 D　计算机常用词汇 ················· 195

附录 E　工科学生学习英语的基本要求 ················· 223

附录 F　部分答案 ················· 225

附录 G　参考译文 ················· 232

参考文献 ················· 265

Unit 1　Computer Structure

Text A　The Function of Computer

1. Central processing unit and microprocessor

A general purpose computer has four main **components**❶: the arithmetic logic unit (ALU), the control unit, the memory, and the input and output devices (collectively termed I/O). These parts **are interconnected by**❷ buses, often **made of**❸ groups of wires.

Inside each of these parts are trillions of small **electrical circuits**❹ which can be turned off or on by means of an electronic switch. Each circuit represents a bit (binary digit) of information so that when the circuit is on it represents a "1", and when off it represents a "0" (in positive logic representation). 注1 The circuits are arranged in logic gates so that one or more of the circuits may control the state of one or more of the other circuits.

The control unit, ALU, registers, and basic I/O (and often other hardware closely linked with these) are collectively known as a central processing unit (CPU). Early CPUs were composed of many separate components but since the mid-1970s CPUs have typically been constructed on a single **integrated circuit**❺ called a microprocessor. 注2

Since 1985, numerous processors implementing some version of the MIPS architecture have been designed and widely used. MIPS (an acronym for Microprocessor without Interlocked Pipeline Stages) is a reduced instruction set computer (RISC) instruction set architecture (ISA) developed by MIPS Technologies (formerly MIPS Computer Systems). The early MIPS architectures were 32-bit called MIPS32, with 64-bit versions added later. The MIPS32 instruction set was developed along side the MIPS64 Instruction Set which includes 64-bit instructions. The MIP32 standard included coprocessor 0 control instructions for the first time. Today, the MIP32 instruction set is the most common MIPS instruction set, compatible with most CPUs. Due to its relative simply, the MIP32 instruction set is also the most common instruction set thought in computer architecture university courses. MIPS32 Add Immediate Instruction is shown in Figure 1-1.

The control unit (often called a control system or central controller) manages the

---

❶ 元件,部件
❷ 由……联系起来
❸ 由……组成
❹ 电路
❺ 集成电路

MIPS32 Add Immediate Instruction

001000	00001	00010	0000000101011110
OP Code	Addr 1	Addr 2	Immediate Value

Equivalent mnemonic: addi &r1, &r2, 350

Figure 1-1 Diagram showing how a particular MIPS architecture instruction would be decoded by the control system

computer's various components; it reads and interprets (**decodes**)❻ the program instructions, **transforming** them **into**❼ a series of control signals which activate other parts of the computer.注3 Control systems in **advanced**❽ computers may change the order of some instructions so as to improve **performance**❾.

A key component common to all CPUs is the program counter: a special memory cell (a register) that keeps track of which location in memory the next instruction is to be read from.

The control system's function is as follows—note that this is a simplified description, and some of these steps may be performed **concurrently**❿ or in a different order depending on the type of CPU:

Read the code for the next instruction from the cell indicated by the program counter.

Decode the numerical code for the instruction into a set of commands or signals for each of the other systems.

Increment⓫ the program counter so it points to the next instruction.

Read whatever data the instruction requires from cells in memory (or perhaps from an input device). The location of this required data is typically stored within the instruction code.

Provide the necessary data to an ALU or **register**⓬.

If the instruction requires an ALU or **specialized**⓭ hardware to complete, instruct the hardware to perform the requested operation.

Write the result from the ALU **back to**⓮ a memory location or to a register or perhaps an output device.

Since the program counter is (conceptually) just another set of memory cells, it can be changed by calculations done in the ALU. Adding 100 to the program counter would cause

---

❻ 解码
❼ 转换
❽ 高级
❾ 性能
❿ 同时发生
⓫ 增量
⓬ 寄存器
⓭ 专门的
⓮ 把……写回到

the next instruction to be read from a place 100 locations further down the program. Instructions that modify the program counter are often known as "jumps" and allow for loops (instructions that are repeated by the computer) and often conditional instruction execution (both examples of control flow).

It is noticeable that the **sequence**⑮ of operations that the control unit goes through to process an instruction is in itself like a short computer program—and indeed, in some more complex CPU designs, there is another yet smaller computer called a microsequencer that runs a microcode program that causes all of these events to happen.

2. Arithmetic/Logic Unit(ALU)

The ALU **is capable of**⑯ performing two classes of operations: **arithmetic**⑰ and logic.

The set of arithmetic operations that a particular ALU supports may be limited to adding and **subtracting**⑱ or might include multiplying or dividing or **trigonometry functions**⑲ (sine, cosine, etc.) and square roots. 注4 Some can only operate on whole numbers (integers) whilst others use floating point to represent real numbers—**albeit with**⑳ limited precision. However, any computer that is capable of performing just the simplest operations can be programmed to **break down**㉑ the more complex operations **into** simple steps that it can perform. 注5 Therefore, any computer can be programmed to perform any arithmetic operation—although it will take more time to do so if its ALU does not directly support the operation. An ALU may also compare numbers and return **boolean**㉒ truth values (true or false) depending on whether one **is equal to**㉓, **greater than**㉔ or **less than**㉕ the other ("is 64 greater than 65?").

Logic operations involve Boolean logic: AND, OR, XOR and NOT. These can be useful both for creating complicated conditional statements and processing Boolean logic.

Superscalar㉖ computers may contain **multiple**㉗ ALUs so that they can process several instructions at the same time. Graphics processors and computers with SIMD and

---

⑮ 一连串,序列
⑯ 能……
⑰ 算术,计算
⑱ 减法
⑲ 三角函数
⑳ 尽管
㉑ 分解成
㉒ 布尔
㉓ 等同于
㉔ 大于
㉕ 小于
㉖ 超标量体系结构
㉗ 多个

MIMD features often provide ALUs that can perform arithmetic on **vectors**㉘ and **matrices**㉙.

3. Memory

Computer data **storage**㉚, often called storage or memory, is a technology consisting of computer components and recording media that are used to retain digital data. Magnetic-core memory shown as Figure 1-2 was the predominant form of random-access computer memory for 20 years between about 1955 and 1975.

Figure 1-2　Magnetic core memory was the computer memory of choice

Magnetic core memory was popular main memory for computers through the 1960s, until it was replaced by semiconductor memory.

A computer's memory can be viewed as a list of cells into which numbers can be placed or read. Each cell has a numbered "address" and can store a single number. <u>The computer can be instructed to "**put** the number 123 **into**㉛ the cell numbered 1357" or to "add the number that is in cell 1357 to the number that is in cell 2468 and put the answer into cell 1595".</u> 注6 The information stored in memory may represent practically anything. Letters, numbers, even computer instructions can be placed into memory with equal ease. Since the CPU does not differentiate between different types of information, it is the software's responsibility to give **significance**㉜ to what the memory sees as **nothing but**㉝ a series of numbers.

In almost all modern computers, each memory cell is set up to store binary numbers in groups of eight bits (called a byte). Each byte is able to represent 256 different numbers ($2^8 = 256$); either from 0 to 255 or -128 to +127. To store larger numbers, several **consecutive**㉞ bytes may be used (typically, two, four or eight). When negative numbers

---

㉘　向量
㉙　matrix 的复数形式,矩阵
㉚　存储
㉛　将……放入
㉜　意义
㉝　只有,只不过
㉞　连续的

are required, they are usually stored in two's complement notation. Other arrangements are possible, but are usually not seen outside of specialized applications or historical **contexts**⑮. A computer can store any kind of information in memory if it can be represented numerically. Modern computers have billions or even trillions of bytes of memory.

The CPU contains a special set of memory cells called registers that can be read and written **too much more rapidly than**⑯ the main memory area. There are typically between two and one hundred registers depending on the type of CPU. Registers are used for the most frequently needed data items to avoid having to access main memory every time data is needed.^{注7} As data is constantly being worked on, reducing the need to access main memory (which is often slow compared to the ALU and control units) greatly increases the computer's speed.

Computer main memory comes in two **principal**⑰ varieties: random-access memory or RAM and read-only memory or ROM. RAM can be read and written to anytime the CPU commands it, but ROM is pre-loaded with data and software that never changes, so the CPU can only read from it.^{注8} ROM is typically used to store the computer's initial start-up instructions. In general, the contents of RAM are erased when the power to the computer is turned off, but ROM retains its data indefinitely. In a PC, the ROM contains a specialized program called the BIOS that orchestrates loading the computer's operating system from the hard disk drive into RAM whenever the computer is turned on or reset. In **embedded**⑱ computers, which frequently do not have disk drives, all of the required software may be stored in ROM. Software stored in ROM is often called **firmware**⑲, because it is notionally more like hardware than software. Flash memory blurs the distinction between ROM and RAM, as it retains its data when turned off but is also rewritable. It is typically much slower than **conventional**⑳ ROM and RAM however, so its use is **restricted**㉑ to applications where high speed is unnecessary.

In more sophisticated computers there may be one or more RAM cache memories which are slower than registers but faster than main memory.^{注9} Generally computers with this sort of **cache**㉒ are designed to move frequently needed data into the cache automatically, often without the need for any **intervention**㉓ on the programmer's part.

- ⑮ 内容
- ⑯ 快得多的……
- ⑰ 重要的,主要的
- ⑱ 嵌入式
- ⑲ 固件
- ⑳ 常规的
- ㉑ 受约束的
- ㉒ 高速缓存
- ㉓ 干涉

4. Input/Output (I/O)

I/O is the means by which a computer exchanges information with the outside world. Devices that provide input or output to the computer are called **peripherals**. ⑭ On a typical personal computer, peripherals include input devices like the keyboard and mouse, and output devices such as the display and printer. 注10 **Hard disk drives**⑮ shown as Figure 1-3, **floppy disk drives**⑯ and **optical disc**⑰ drives serve as both input and output devices. Computer networking is another form of I/O.

Figure 1-3　Hard disk drives are common storage devices used with computers

Often, I/O devices are complex computers in their own right with their own CPU and memory. A graphics processing unit might contain fifty or more tiny computers that perform the calculations necessary to display 3D graphics. Modern desktop computers contain many smaller computers that **assist**⑱ the main CPU in performing I/O.

5. Multitasking

While a computer may **be viewed as**⑲ running one gigantic program stored in its main memory, in some systems it is necessary to give the appearance of running several programs **simultaneously**⑳. This is achieved by multitasking, i.e. having the computer switch rapidly between running each program **in turn**㉑.

One means by which this is done is with a special signal called an **interrupt**㉒ which can periodically cause the computer to stop executing instructions where it was and do something else instead. 注11 By remembering where it was executing prior to the interrupt, the computer can return to that task later. The interrupt generator might be causing several

---

- ⑭ 外围设备
- ⑮ 硬盘驱动器
- ⑯ 软盘驱动器
- ⑰ 光盘
- ⑱ 帮助,辅助
- ⑲ 看成
- ⑳ 同时
- ㉑ 依次
- ㉒ 打断,中断

hundred interrupts per second, causing a program switch each time, if several programs are running "at the same time". Since modern computers typically execute instructions **several orders of**㊽ magnitude **faster than**㊾ human perception, it may appear that many programs are running at the same time even though only one is ever executing in any given instant. This method of multitasking is sometimes termed "time-sharing" since each program is **allocated**㊿ a "**slice**"㊱ of time in turn.

Before the era of cheap computers, the principal use for multitasking was to allow many people to share the same computer.

Seemingly, multitasking would cause a computer that is switching between several programs to run more slowly—**in direct proportion to**㊲ the number of programs it is running. However, most programs spend much of their time waiting for slow input/output devices to complete their tasks. If a program is waiting for the user to click on the mouse or press a key on the keyboard, then it will not take a "time slice" until the event it is waiting for has occurred. This frees **up time**㊳ for other programs to execute so that many programs may be run at the same time without unacceptable speed **loss**㊴.

6. Multiprocessing

Some computers are designed to **distribute**㊵ their work across several CPUs in a multiprocessing **configuration**㊶, a technique once **employed**㊷ only in large and powerful machines such as supercomputers, **mainframe**㊸ computers and servers. 注12 Multiprocessor and multi-core (multiple CPUs on a single integrated circuit) personal and laptop computers are now widely available, and are being increasingly used in lower-end markets as a result shown as Figure 1-4.

Supercomputers in particular often have highly unique architectures that differ significantly from the basic stored-program architecture and from general purpose computers. They often feature thousands of CPUs, customized high-speed interconnects, and specialized computing hardware. 注13 Such designs tend to be useful only for specialized tasks due to the large scale of program organization required to successfully utilize most of the available

---

㊽ 几个数量级
㊾ 更快的
㊿ 分配
㊱ 片
㊲ 与……成正比
㊳ 节省了时间
㊴ 损失
㊵ 分配
㊶ 配置
㊷ 使用
㊸ 主机,大型机

· 7 ·

resources at once. Supercomputers usually see usage in large-scale **simulation**㉔, graphics rendering, and cryptography applications, **as well as**㉕ with other **so-called**㉖ "**embarrassingly**㉗ parallel" tasks. 注14

Figure 1-4 Cray designed many supercomputers that used multiprocessing heavily

7. Networking and the Internet

Computers have been used to **coordinate**㉘ information between multiple locations since the 1950s. The U. S. military's SAGE system was the first **large-scale**㉙ example of such a system, which **led to a number of**㉚ special-purpose **commercial**㉛ systems like **Sabre**㉜.

In the 1970s, computer engineers at research institutions throughout the United States began to link their computers together using telecommunications technology. 注15 This effort was funded by ARPA (now DARPA), and the computer network that it produced was called the ARPANET. The technologies that made the **ARPANET**㉝ possible spread and evolved.

In time㉞, the network spread beyond academic and military institutions and became known as the Internet shown as Figure 1-5. The emergence of networking involved a **redefinition**㉟ of the nature and boundaries of the computer. Computer operating systems and applications were modified to include the ability to define and access the resources of other computers on the network, such as peripheral devices, stored information, and the like,

---

㉔ 模仿
㉕ 也
㉖ 称为
㉗ 使人为难地
㉘ 使协调
㉙ 大规模地
㉚ 导致许多的……
㉛ 商业的
㉜ 商业系统名称
㉝ 互联网的始祖
㉞ 最后
㉟ 补救

as extensions of[70] the resources of an individual computer.[注16] Initially these facilities were available primarily to people working in high-tech environments, but in the 1990s the spread of applications like E-mail and the World Wide Web, **combined with**[71] the development of cheap, fast networking technologies like Ethernet and ADSL saw computer networking become almost ubiquitous. In fact, the number of computers that are networked is growing phenomenally. A very large proportion of personal computers regularly connect to the Internet to communicate and receive information. "**Wireless**"[72] networking, often **utilizing**[73] mobile phone networks, has meant networking is becoming increasingly ubiquitous even in mobile computing environments.[注17]

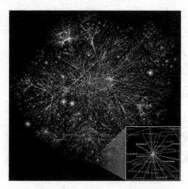

Figure 1-5　Visualization of a portion of the routes on the Internet

Words

albeit	*conj.*（连词）	即使
allocate	*vt.*（动词）	分配
arithmetic	*n.*（名词）	算术,计算
ARPAnet	*n.*（名词）	互联网的始祖
assist	*vt. & vi.*（动词）	帮助,辅助
Boolean	*adj.*（形容词）	布尔
cache	*n.*（名词）	高速缓存
commercial	*adj.*（形容词）	商业的,商务的
component	*n.*（名词）	元件部分
concurrently	*adv.*（副词）	兼;同时发生地
configuration	*n.*（名词）	构造,配置

---

⑯ 作为扩展
⑰ 和……一起
⑱ 无线的
⑲ 应用,使用

consecutive	*adj.*（形容词）	连续的
context	*n.*（名词）	内容
conventional	*adj.*（形容词）	常规的，非核的
coordinate	*vt.*（动词）	使协调；使调和
decode	*vt.*（动词）	解码
distribute	*vt.*（动词）	分配，分给
electrical circuit	*compound noun*（复合名词）	电路
embarrassingly	*adv.*（副词）	使人为难地
embeded	*adj.*（形容词）	嵌入的
employ	*vt.*（动词）	使用，利用
firmware	*n.*（名词）	固件
floppy disk drives	*compound noun*（复合名词）	软盘驱动
hard disk drives	*compound noun*（复合名词）	硬盘驱动
increment	*n.*（名词）	增加，增量
integrated circuit	*compound noun*（复合名词）	集成电路
interrupt	*vt.*（动词）	打断，中断
intervention	*n.*（名词）	干涉
loss	*n.*（名词）	损失
mainframe	*n.*（名词）	主机
matrices	*n.*（名词）	matrix 的复数形式，矩阵
multiple	*adj.*（形容词）	多个
multitasking	*n.*（名词）	多(重)任务处理
optical disc	*compound noun*（复合名词）	光盘
performance	*n.*（名词）	执行，性能，表现
peripheral	*n.*（名词）	外围设备
principal	*adj.*（形容词）	重要的，主要的
redefinition	*n.*（名词）	赎回；补救；兑现
register	*n.*（名词）	寄存器
restrict	*adj.*（形容词）	有限的，受约束的
Sabre	*n.*（名词）	商业系统名称
scale	*n.*（名词）	比例
sequence	*n.*（名词）	一连串
significance	*n.*（名词）	意义
simulation	*n.*（名词）	模仿；模拟
simultaneously	*adv.*（副词）	同时地
slice	*n.*（名词）	片
specialize	*adj.*（形容词）	专门的，专科的
storage	*n.*（名词）	存储

subtract	*vt.* & *vi.* (动词)	减去
superscalar	*n.* (名词)	超标量体系结构
transform	*vt.* & *vi.* (动词)	转换
trigonometry function	*compound noun* (复合名词)	三角函数
utilize	*vt.* (动词)	应用,使用,利用
vector	*n.* (名词)	向量
wireless	*n.* (名词)	无线电

Phrases

albeit with	尽管
as extensions of	为扩展的
as well as	也;和……一样;不但……而且
be capable of	能够
be viewed as	被……看成/看作
break down into	分解成……
combined with	化合,联合;连同
faster than	比……快
greater than	[数]大于
in direct proportion to	成正比
in time	及时;适时
in turn	轮流,依次
interconnected by	由……联系起来
be equal to	等于
large-scale	大规模地
led to a number of	导致了许多
less than	小于
made of	由……组成,由……构成
nothing but	只有;只不过
put into	把……放进(或关进、投入)
several orders of	几个数量级
so-called	所谓的;号称的
too much more rapidly than	更迅速地超过
transforming into	转变成
up time	节省了时间
write back to	回复,回信,把……写回到

Exercises

【Ex1】 Answer the questions according to the text:

(1) How many components does a general purpose computer have? And what are they?
(2) What is a central processing unit?
(3) What is the control unit's main task?
(4) What are the ALU operations?
(5) What is I/O? Can you list some input devices?

【Ex2】 Translate into Chinese:

(1) Inside each of these parts are trillions of small electrical circuits which can be turned off or on by means of an electronic switch. Each circuit represents a bit (binary digit) of information so that when the circuit is on it represents a "1", and when off it represents a "0" (in positive logic representation).

(2) Adding 100 to the program counter would cause the next instruction to be read from a place 100 locations further down the program. Instructions that modify the program counter are often known as "jumps" and allow for loops (instructions that are repeated by the computer) and often conditional instruction execution (both examples of control flow).

(3) The set of arithmetic operations that a particular ALU supports may be limited to adding and subtracting or might include multiplying or dividing or trigonometry functions (sine, cosine, etc.) and square roots. Some can only operate on whole numbers (integers) whilst others use floating point to represent real numbers—albeit with limited precision.

(4) Computer main memory comes in two principal varieties: random-access memory or RAM and read-only memory or ROM.

(5) While a computer may be viewed as running one gigantic program stored in its main memory, in some systems it is necessary to give the appearance of running several programs simultaneously.

【Ex3】 Choose the best answer:

(1) Some computers are designed to _____ their work across several CPUs in a multiprocessing configuration.
  A. distribute   B. distributed
  C. distributing   D. distributes

(2) One means by which this is done is with a special signal called an _____ which can periodically cause the computer to stop executing instructions where it was and do something else instead.

A. pause　　　　B. stop　　　　C. interrupt　　　　D. exit

（3）Hard disk drives, floppy disk drives and _____ drives serve as both input and output devices.

　　A. Optical magnetic　　　　B. keyboard
　　C. mouse　　　　　　　　　D. optical disc

（4）In almost all modern computers, each memory cell is set up to store binary numbers in groups of _____（called a byte）. Each byte is able to represent 256 different numbers（$2^8=256$）; either from 0 to 255 or -128 to $+127$.

　　A. One bit　　　B. two bits.　　　C. eight bits　　　D. eight bit

（5）A general purpose computer has four main components: the arithmetic logic unit（ALU）, _____, the memory, and the input and output devices（collectively termed I/O）.

　　A. the control unit　　　　B. mouse
　　C. display　　　　　　　　D. disc

批　注

注1　主语: Each circuit; 谓语: represents; so that 引导了一个目的状语从句; when the circuit is on 是目的状语从句的时间状语从句; it 为代词; and when off 与 when the circuit is on 是并列的时间状语。

注2　主语: Early CPUs; 谓语: were composed of, have typically been constructed, 通过 but 把两句话并列; called a microprocessor 后置定语修饰 integrated circuit。

注3　主语: The control unit; 谓语: manages。it 是代词, 指 control unit; transforming 分词引导独立结构; which activate other parts of the computer 中 which 的先行词为 control signals。

注4　主语: the set of arithmetic operations。that 引导一个定语从句; particular ALU 是从句主语; supports 是从句谓语; may be limited to 与 might 为并列结构。

注5　主语: any computer; 谓语: can be programmed。定语从句: that is capable of performing just the simplest operations; programmed... into: 将……编程转换为。

注6　主语: computer; 谓语: can be instructed to。put the number 123 into the cell numbered 1357 与 add the number that is in cell 1357 to the number 为并列结构; that is in cell 2468, that 的先行词为 number。

注7　depending on the type of CPU 分词短语作定语, 修饰 registers; Registers 为主语; are used 为谓语; data is needed 前面省略了 when, 为时间定语从句。

注8　random-access memory and read-only memory or ROM 是 two principal varieties 的同位语; 句子主语: computer main memory; so the CPU can only read from it 为结果状语从句。

注9　In more sophisticated computers 为状语; which are slower than registers 中 which 的先行词为 RAM cache memories; but faster than main memory 省略了 which are。

注10　that provide input or output to the computer 中 that 的先行词为 Devices; 句子主谓语: devices are called. 这句话 input device like the keyboard and mouse, and output devices such as the display and printer 为并列的关系, 都属于 peripherals。

注11　by which this is done 是一个定语从句, 修饰 one means; which can periodically cause the computer to stop executing instruction 中 which 的先行词是 interrupt, 引导一个定语从句。

注12　主语: Some computers; 谓语: are designed to。a technique once employed only in large and powerful machines such as supercomputers, mainframe computers and servers 这句省略了 which is, 为非限制性定语从句。

注 13　customized high-speed interconnects 前面省略了 They often; and specialized computing hardware 前面省略了 They often。

注 14　graphics rendering 前面省略了 supercomputers usually see usage in; cryptography（密码学）applications 前面省略了 supercomputers usually see usage in。

注 15　In the 1970s 为时间短语; computer engineers 为句子主语; began to 为谓语; at research institutions 是定语; throughout the United States 是定语; link ... together 意为: 把……相连接; using telecommunications technology 为方式状语, 意思是: 通过使用……。

注 16　Computer operating systems and applications 为主语; were modified to: 被动语态, 被适应为……; include... and: 包含; and the like: 等等; as extensions of: 作为……的扩展。

注 17　"Wireless" networking 为句子的主语; often utilizing mobile phone networks 对前面 wireless networking 补充说明; 谓语: has meant; networking is becoming... 这句前面省略了 that。

Text B　The Future of Computer Technology

Over time, people have always wanted the fastest computer they could get, right?

When the 386 came out, everyone thought it was **blazing**❶ fast, and needed one.

Suddenly, **apps**❷ were written that used that power and the 386 became slow, and the 486 came out, and everyone needed one.

The trend is that we use hardware performance as it becomes available, and then want more. Kind of like the "never enough money" argument.

But the **assumption**❸ that this trend will continue is **flawed**❹.

The fact is your average Joe only needs a computer to do so much. Basically, the computer as an "**appliance**"❺ **is able to**❻ fully **interact with**❼ our senses, which are limited. This means, until we see the **smell-o-mouse**❽, we essentially need:

Full speed video output:
　　Movie decoding/presentation/generation, gaming;

---

❶ 酷热的
❷ applications 的缩写
❸ 假定, 臆断
❹ 有缺陷的
❺ 器械
❻ 能够
❼ 交互
❽ 嗅觉鼠标

Full speed video input:

　　Vision recognition, movie saving/encoding/compressing;

Full speed audio output:

　　Sound decoding/mixing, voice synthesis;

Full speed audio input:

　　Voice recognition, recording/compressing;

　　Realize this argument applies to the "Average Joe"-not computer **geeks**❾ like you and I who will want to run huge simulations and do software development or CAD, and will still need to have a full-powered workstation at home.^{注1} We are the **minority**❿.

　　Most of these requirements are currently available with today's computers. Many of them are also available, just not at full speed yet (such as full video generation, 'flawless' voice recognition) and will be available with **off-the-shelf**⓫ computing equipment in a matter of years.

　　Have you noticed that people don't talk as much about processor speed? If you go into your local computer store, they talk these days just as much about RAM, disk, screen size, battery and other features. It was only a few years ago where processor speed was king, and now it's becoming less **vital**⓬, and soon it will only be a **footnote**⓭ to the specs that a computer will **advertise**⓮.

　　So then what happens?

　　Things are going to get much smaller, and fast.

　　It used to be that a laptop was a bulky and slow and weak machine.^{注2} It couldn't do all the tasks of a home computer. This isn't true anymore, my 2 pound ultra-slim laptop (which I paid $150 for; incidentally), can basically keep up with my **state-of-the-art**⓯ home workstation.^{注3} You can already see the emergence of more and more powerful small computers. Things like Palm are only the **temporary**⓰ off shoots of a new direction in business. Palm is a **subset**⓱ of a real OS, and a subset of a real computer. Pretty soon you'll be able to get a full computer in the same size as a Palm, and it will cost about the same. Which would you rather have, the personal organizer that has a built-in micro

---

❾ 怪杰
❿ 少数
⓫ 现成的
⓬ 极重要的,必不可少的
⓭ 脚注
⓮ 公布,宣传
⓯ 最高级的
⓰ 临时的,暂时的
⓱ 子集

browser, or the personal organizer that you just downloaded the latest **Netscape**⑱/**Mozilla**⑲/**Opera**⑳ into with the newest **version**㉑ of flash/java/shockwave?注4 The **personal organizer**㉒ that has a provided address list app. or one that can run any address list app you want, and play **Doom**㉓?

This whole argument applies to data storage as well.

Everyone wants the biggest hard drive/RAM they can get. We keep coming up with more ways to fill them up (mp3s, video...), and we need to keep upgrading. Although these days, running out of space on your hard drive is a much rarer occurrence. I used to have to **juggle**㉔ things between compressing files and moving them to floppy disk storage.

These days I don't even own any disks. And I only write CDs generally as a backup method for information that is also on my computer.

In less than 10 years you will be able to store full videos on your hard drive just like you can store audio on there today. You'll have a movie playlist with 1000's of movies, just like your mp3 player today.

And a few years later you'll be able to put this on the end of your **keychain**㉕. Are you really going to need to get a bigger hard drive after that? At some point you'll be able to store more video and audio than you'll ever be able to **experience in**㉖ your **continuous lifetime**㉗. At that point, you won't need the next drive size, now will you?

I admit there's one big difference between my laptop and my home computer. And that's the keyboard, the 21" monitor and the CD-RW. (Though most **laptops**㉘ have DVD/CD-RW these days...)

Now let's look at the future.

But before doing that, consider the past, consider what has happened with the computer industry.

Initially the money was in mainframes. Personal computers existed as a sort of **clunky hobby**㉙, but the bucks were flowing **in the mainframe industry**㉚. And that has

---

⑱ 美国 Netscape 公司,以开发 Internet 浏览器而闻名
⑲ Mozilla 公司 Firefox 是该公司产品
⑳ Opera 软件公司制造的一款浏览器
㉑ 版本
㉒ 个人信息管理器
㉓ "毁灭战士"游戏
㉔ 歪曲,颠倒
㉕ 钥匙链
㉖ 经历
㉗ 一生
㉘ 笔记本计算机
㉙ 笨拙的业余爱好
㉚ 大型机主机行业

everything to do with how useful a PC could be compared to a mainframe. As performance increased, servers started to do lots of things that mainframes could do, and the $ $ was in the **mainframes**❶ market. The real issue is that there is a certain amount of computing power (per dollar) that actually is critically useful. Anything less than that and it'll be too slow, and you'll want more speed. Anything more powerful than that and it'll only be useful **on the fringe**❷, in the technical and development markets. Then it was workstations, and they started to reach this quantum of computing and started replacing servers. Then it was PCs. The PC market took off and PCs started to replace workstations. This **watermark**❸ was best publicized in the movie industry when movies like Titanic were advertised to have been created for less money by using huge PC farms instead of workstation farms. 注5 Now we're moving towards laptops, and then subnotebooks will be next, then handhelds. Finish the progression yourself.

So here's my **forecast**❹ for the future.

You'll have a computing unit. It will be a processor that is xxx MHz, where 'xxx' is "fast enough." It will also have solid state storage that is xxx GB, where 'xxx' is "big enough." And this little guy will be the size of a cell phone or smaller, and cost about the same.

And you'll go home and drop it into a port that **hooks**❺ it **up** to a monitor and keyboard. And on the road you'll put it in a shell with a Palm sized touch display that lets you access all of your info, and even listen to music or watch a video besides. And then you might **pop** it **into**❻ your camera shell (or your organizer shell might have a lens on it), and you can take as many pictures as you want, because your storage space can hold 1000's of movies, so you'll never fill it up with stills/home video, and you'll never have to sync with your home machine. 注6 You may even have full wireless, and your photos can automatically go online to your web photo **album**❼. When you get home from your trip, you can drop the unit into your theater system and you can watch movies or see **footage from your trip**❽. You could put some form of net-currency **encrypted on**❾ your unit, so you can use it to purchase things as well. You could put it in your car when you are driving and listen to every song you own. You could have a GPS in your car and it could use a

---

❶ 大型主机
❷ 边缘
❸ 水印
❹ 预见
❺ 钩起
❻ 顺道进入
❼ 相册
❽ 从你的旅行镜头开始看起
❾ 加密

service like **Map Quest**⑩ to give you directions. And your sweetie will send you email and you'll use a headset peripheral and telephony software to call her back.注7

But the important point is that you won't need to get multiple computing units, and you won't need to have a home computer anymore. You'll have one computing unit, and it will go with you like a wallet, because it will replace your credit cards, your cell phone, your fax machine, your DVD player, your stereo, your browser, your **organizer**⑪, your email, etc..注8

And then there will be some museum where all these towers and cases sit, and people can go look at them and chuckle just like we laugh at ENIAC and the **vacuum tube**⑫ today.

Words

advertise	vt. & vi.（动词）	公布,宣传
album	n.（名词）	粘贴簿,集邮簿,相册
appliance	n.（名词）	器具,器械,装置
apps	n.（名词）	applications 的缩写
blazing	adj.（形容词）	酷热的;炽热的
doom	n.（名词）	厄运;死亡;判决;世界末日
flawed	adj.（形容词）	有裂纹的,有瑕疵的;有缺陷的
footnote	n.（名词）	脚注
forecast	v.（动词）	预见
geek	n.（名词）	怪杰
hook	vt. & vi.（动词）	钩住,吊住,挂住
juggle	n.（名词）	歪曲,颠倒
keychain	n.（名词）	钥匙链
laptop	n.（名词）	掌上计算机
mainframe	n.（名词）	主机
mapquest	vt. & vi.（动词）	地图定位
minority	adj.（形容词）	少数,小部分
Mozilla	n.（名词）	Mozilla 公司开发的一款浏览器
Netscape	n.（名词）	美国 Netscape 公司开发的浏览器
Opera	n.（名词）	Opera 软件公司制造的一款浏览器
organizer	n.（名词）	管理器、助理器
subset	n.（名词）	子集

⑩ 地图定位
⑪ 管理器、助理器
⑫ 真空管,电子管

temporary	*adj.*（形容词）	临时的,暂时的
version	*n.*（名词）	版本,形式
vital	*adj.*（形容词）	极重要的,必不可少的
watermark	*n.*（名词）	水印

Phrases

be able to	能够
clunky hobby	笨拙的业余爱好
continuous lifetime	持续一生
encrypted on	加密
experience in	有经验；有……的经验
footage from your trip	从你的旅行镜头开始
from your trip	从你的旅行开始
in the mainframe industry	大型机主机行业
interact with	与……相互作用
off-the-shelf	现成的；常备的；成品的
on the fringe	边缘
personal organizer	个人信息管理器
pop…into	匆匆地走进……
smell-o-mouse	嗅觉鼠标
state-of-the-art	最先进的
vacuum tube	真空管,电子管

批 注

注 1　Average Joe 与 geeks 对应,意为平常人与怪杰；not computer geeks 这里作为插入语；who 引导一个定语从句,先行词为 you and I；run 与 do need 并列。do software development：从事软件开发工作。

注 2　used to be：过去常常；that 引导一个宾语从句；a laptop 是从句主语；a bulky, slow and weak machine 为表语。

注 3　laptop 为主语；can keep up with 为谓语；my 2 pound ultra-slim 作定语。

注 4　the personal organizer that has a built-in micro browser 与 the personal organizer that you just downloaded 为并列结构,各有 that 引导一个定语从句,downloaded into：将最新的 flash/java/shockwave 装载到 Netscape/Mozilla /Opera。

注 5　when 引导一个时间状语从句；instead of workstation farms 省略了 using。

注 6　you can take as many pictures as you want 这是一个比较句,与……一样多；sync with：与……同步。

注 7　由三个简单并列句组成；sweetie 表示甜心；headset peripheral：戴在头上的耳机或听筒。

注 8　because 引导一个原因状语从句；your credit cards, your cell phone, your fax machine, your DVD player, your stereo, your browser, your organizer, your email 为并列结构,作 replace 的宾语。

Associated Reading

Brief Introduction of Motherboard

Motherboard is one of the main components of a computer and made of the printed circuit board (PCB). The motherboard is also known as mainboard, system board, logic board or sometimes, shortened as "mobo". It serves as the backbone of a system because it provides all the electrical connections by which other components of the system communicate. Aside from that, it hosts the central processing unit (CPU), and other devices. This is illustrated in Figure 1-6.

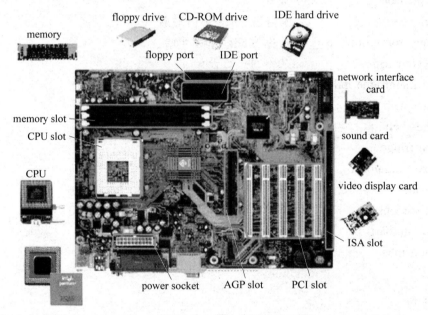

Figure 1-6　The picture of motherboard

批 注

(1) memory: 内存
(2) memory slot: 内存插槽
(3) CPU slot: CPU 插槽
(4) CPU: 中央处理器
(5) floppy drive: 软盘驱动器
(6) floppy port: 软驱接口
(7) CD-ROM drive: 光盘驱动器
(8) IDE port: IDE 接口
(9) network interface card: 网卡
(10) sound card: 声卡

(11) video display card：显示卡
(12) ISA slot：ISA 插槽
(13) PCI slot：PCI 插槽
(14) AGP slot：AGP 插槽
(15) power socket：电源插座

Unit 2　Programming Language

Text A　Functions in C

Almost all programming languages have some **equivalent**❶ of the function. **You may have met**❷ them under the **alternative**❸ names **subroutine**❹ or procedure. 注1

Some languages **distinguish between**❺ functions which return variables and those which don't. 注2 C assumes that every function will return a value. If the programmer wants a return value, this is achieved using the return statement. If no return value is required, none should be used when calling the function.

Here is a function which raises a double to the power of an unsigned, and returns the result.

```
double power(double val, unsigned pow)
{       double ret_val = 1.0;
        unsigned i;

        for(i = 0; i < pow; i++)
            ret_val *= val;

        return(ret_val);
}
```

The function follows a simple **algorithm**❻, multiplying the value by itself **pow**❼ times. A for loop is used to control the number of multiplications, and **variable**❽ ret_val stores the value to be returned. 注3 Careful programming has ensured that the **boundary condition**❾ is correct too.

Let us examine the details of this function:

```
double power (double val, unsigned pow)
```

---

❶ 等价,对等的
❷ 遇见
❸ 选择,替代的
❹ 子程序
❺ 区分
❻ 算法
❼ 乘幂
❽ 变量
❾ 边界条件

This line begins the function definition. It tells us the type of the return value, the name of the function, and a list of arguments used by the function. The arguments and their types are enclosed **in brackets**❿, each pair separated by commas.

The body of the function is bounded by **a set of curly brackets**⓫. Any variables declared here will be treated as local unless specifically **declared as**⓬ static or extern types.

```
return(ret_val);
```

On reaching a return statement, control of the program returns to the calling function. The bracketed value is the value which is returned from the function. If the final closing curly bracket is reached before any return value, then the function will return automatically, any return value will then be meaningless. 注4

The example function can be called by a line in another function which looks like this:

```
result=power(val, pow);
```

This calls the function power assigning the return value to variable result.

Here is an example of a function which does not return a value.

```
void error_line(int line)
{    fprintf (stderr, "Error in input data: line %d\n", line);
}
```

The definition uses type void which is optional. 注5 It shows that no return value is used. Otherwise the function is much **the same as**⓭ the previous example, except that there is no return statement. 注6 Some void type functions might use return, but only to force an early exit from the function, and not to return any value. This is **rather like**⓮ using break to jump out of a loop. 注7

This function also demonstrates a new feature.

```
fprintf (stderr, "Error in input data: line %d\n", line);
```

This is a variant on the printf statement; fprintf sends its output into a file. In this case, the file is stderr. stderr is a special UNIX file which serves as the channel for error messages. It is usually connected to the console of the computer system, so this is a good way to display error messages from your programs. 注8 Messages sent to stderr will appear on screen even if the normal output of the program has been redirected to a file or a printer. 注9

The function would be called as follows:

---

❿　括号内
⓫　大括号
⓬　声明
⓭　相同
⓮　更像

```
error_line(line_number);
```

Words

algorithm	*n.*（名词）	［计］［数］算法,运算法则
alternative	*adj.*（形容词）	供选择的;选择性的;交替的
equivalent	*adj.*（形容词）	等价的,相等的;同意义的
pow	*n.*（名词）	乘幂
subroutine	*n.*（名词）	子程序
variable	*n.*（名词）	变量;可变物,可变因素

Phrases

a set of curly brackets	大括号
boundary condition	［数］边界条件,界面条件
declared as	声明
distinguish between	区别;分辨
in brackets	括号内
rather like	更像
the same as	相同

Exercises

【Ex1】 Answer the questions according to the text：

(1) What is C? What does that mean?

(2) What is the file stderr for?

(3) What is the boundary of a function body?

(4) What are the standard types for variables in C?

(5) What is the way your computer remembers things?

【Ex2】 Translate into Chinese：

(1) Careful programming has ensured that the boundary condition is correct too.

(2) The arguments and their types are enclosed in brackets, each pair separated by commas.

(3) The example function can be called by a line in another function which looks like this.

(4) The body of the function is bounded by a set of curly brackets.

(5) This is a variant on the printf statement, fprintf sends its output into a file.

(6) Stderr is a special UNIX file which serves as the channel for error messages.

(7) Messages sent to stderr will appear on screen even if the normal output of the program has been redirected to a file or a printer.

(8) This calls the function power assigning the return value to variable result.

【Ex3】 Choose the best answer:

(1) In C Language, a _____ is a series of characters enclosed in double quotes.

  A. matrix   B. string   C. program   D. stream

(2) In C Language, _____ are used to create variables and are grouped at the top of a program block.

  A. declarations   B. dimensions   C. comments   D. descriptions

(3) A _____ consists of the symbols, characters, and usage rules that permit people to communicate with computer.

  A. programming language   B. network

  C. storage   D. function

(4) A(An) _____ software, also called end-user program, includes database programs, word processors, spreadsheets etc.

  A. application   B. system   C. compiler   D. utility

(5) If no return value is required, _____ should be used when calling the function.

  A. a value   B. some value   C. none   D. no Frequency

批 注

注1　under the alternative names：称呼为；主语：you；谓语：have met。or procedure 前省略了 under the name,此句的含义：你可能遇到过称为子程序或进程的函数。

注2　which 引导一个定语从句；those 是指 functions；which don't 省略了 return variables。

注3　主语：A for loop,其含义是指一个 for 循环；multiplications 这个单词为复合词,其含义是指多个应用程序。此句的含义是：一个 for 语句用来控制一些复合应用程序,而变量 ret_val 用来存储返回的值。

注4　if 引导一个条件状语从句。如果在返回值前有一个结束的大括号,那么这个函数将自动返回一个值,后面写的任何返回值都无意义。

注5　此句话的含义是指这个定义使用的是 void 类型,这个属性是可选的。

注6　否则这个函数将与前例一样,除了含有一个返回语句不同之外。

注7　这有点类似于使用 break 语句跳出 loop 循环。

注8　它通常与控制台连接,这样可以更好地把程序中的错误显示出来。

注9　主语：Messages sent to stderr；谓语：will appear；even if 引导一个条件状语从句；redirected 是指重定向,这句话的含义是把消息传送给 stderr,通过 stderr 把消息显示在屏幕上,即常见的程序输出是把程序重定向至一个文件或打印机。

Text B J2ME Related

1. How J2ME is organized

Traditional computing devices use fairly standard hardware configurations such as a display, keyboard, mouse, and large amounts of memory and permanent storage.

However, the new **breed**❶ of computing devices lacks hardware configuration **continuity**❷ among devices. Some devices don't have a display, permanent storage, keyboard, or mouse. And **memory availability**❸ is inconsistent among small computing devices.

The lack of uniform hardware configuration among the small computing devices poses a **formidable**❹ challenge for the Java Community Process Program, which is **charged with**❺ developing standards for the JVM and the J2ME for small computing devices.

J2ME must service many different kinds of small computing devices, including screen phones, digital **set-top boxes** ❻ used for cable television, cell phones, and personal digital assistants. 注1 The challenge for the **Java Community Process Program**❼ is to develop a Java standard that can be implemented on small computing devices that have nonstandard hardware configurations.

The Java Community Process Program has used a twofold approach to addressing.

The needs of small computing devices. First, they defined the Java run-time environment and core classes that operate on each device. This is referred to as the *configuration*. A configuration defines the Java Virtual Machine for a particular small computing device. There are two configurations, one for **handheld**❽ devices and the other for plug-in devices. Next, the Java Community Process Program defined a profile for categories of small computing devices. 注2 A *profile* consists of classes that enable developers to implement features found on a related group of small computing devices.

2. J2ME configurations

There are two configurations for J2ME as of this writing. These are Connected Limited Device Configuration (CLDC) and the Connected Device Configuration (CDC). The CLDC

---

❶ 品种,类型
❷ 延续性,一系列
❸ 内存容量
❹ 难以克服的,巨大的
❺ 承担,负责
❻ 机顶盒
❼ Java 社团处理程序
❽ 手持式

is designed for 16-bit or 32-bit small computing devices with limited amounts of memory.

CLDC devices usually have between 160KB and 512KB of available memory and are battery powered. They also use an inconsistent, small-bandwidth network **wireless**⑨ connection and may not have a user interface. CLDC devices use the K-Java Virtual Machine (KVM) implementation, which is a **stripped-down**⑩ version of the JVM. CLDC devices include pagers, personal digital assistants, cell phones, **dedicated**⑪ **terminals**⑫, and handheld consumer devices with between 128KB and 512KB of memory.^{注3} CDC devices use a 32-bit architecture, have at least two megabytes of memory available, and implement a complete functional JVM. CDC devices include digital set-top boxes, home appliances, **navigation**⑬ systems, **point-of-sale terminals**⑭, and smart phones.

3. J2ME profile

A profile consists of Java classes that enable implementation of features for **either** a particular small computing device **or**⑮ for a class of small computing devices. Small computing technology continues to **evolve**⑯, and with that, there is an **ongoing**⑰ process of defining J2ME profiles. Seven profiles have been defined as of this writing. These are the **Foundation Profile**⑱, **Game Profile**⑲, **Mobile Information Device Profile**⑳, **PDA Profile**㉑, **Personal Profile**㉒, **Personal Basis Profile**㉓, **and RMI Profile**㉔.

The Foundation Profile is used with the CDC configuration and is the **core for**㉕ nearly all other profiles used with the CDC configuration because the Foundation Profile contains core Java classes.

- The Game Profile is also **used with**㉖ the CDC configuration and contains the necessary classes for developing game applications for any small computing device

---

⑨ 无线的
⑩ 简装的
⑪ 专用于
⑫ 终端
⑬ 导航
⑭ 销售点终端机
⑮ 或……或
⑯ 逐步发展
⑰ 即将
⑱ 基础配置文件
⑲ 游戏配置文件
⑳ 移动信息装备配置文件
㉑ 个人助理配置文件
㉒ 个人配置文件
㉓ 个人基本信息配置文件
㉔ RMI 配置文件
㉕ 核心
㉖ 使用

that uses the CDC configuration.
- The Mobile Information Device Profile (MIDP) is used with the CLDC configuration and contains classes that provide local **storage**❷⓻, a user interface, and networking **capabilities**❷⓼ to an application that runs on a mobile computing device such as Palm OS devices. 注4 MIDP is used with wireless Java applications.
- The PDA Profile (PDAP) is used with the CLDC configuration and contains classes that **utilize**❷⓽ sophisticated resources(14) found on personal digital assistants. These features include better displays and larger memory than similar resources found on MIDP mobile devices (such as cell phones). 注5
- The Personal Profile is used with the CDC configuration and the Foundation Profile and contains classes to implement a complex user interface. The Foundation Profile provides core classes, and the Personal Profiles provide classes to implement a sophisticated user interface, which is a user interface that is capable of displaying multiple windows at a time.
- The Personal Basis Profile is similar to the Personal Profile in that it is used with the CDC configuration and the Foundation Profile. However, the Personal Basis Profile provides classes to implement a simple user interface, which is a user interface that is capable of displaying one window at a time.
- The RMI Profile is used with the CDC configuration and the Foundation Profile to provide Remote Method Invocation classes to the core classes contained in the Foundation Profile.

There will likely be many profiles as the proliferation of small computing devices continues. Industry groups within the Java Community Process Program (java. sun. com/aboutjava/communityprocess) define profiles. Each group establishes the standard profile used by small computing devices manufactured by that industry. A CDC profile is defined by **expanding**❸⓪ upon core Java classes found in the Foundation Profile with classes specifically targeted to a class of small computing device. These device-specific classes are contained in a new profile that enables developers to create **industrial-strength**❸⓵ applications for those devices. However, if the Foundation Profile is specific to CDC, not all profiles are expanded upon the core classes found in the Foundation Profile.

Keep in mind that applications can access a small computing device's software and hardware features only if the necessary classes to do so are contained in the JVM and in the profile used by the developer 注6.

---

❷⓻ 存储
❷⓼ 能力
❷⓽ 利用
❸⓪ 扩展
❸⓵ 工业级强度

Words

breed	n.（名词）	［生物］品种；种类，类型
capabilities	n.（名词）	能力(capability 的复数)；功能；性能
continuity	n.（名词）	连续性；一连串；分镜头剧本
dedicate	adj.（形容词）	专用的；专注的；献身的
evolve	vt. & vi.（动词）	发展，进化；使逐步形成；推断出
expand	vt. & vi.（动词）	扩大，扩展(expand 的现在分词形式)；使膨胀，详述
formidable	adj.（形容词）	可怕的；令人敬畏的；艰难的；强大的
handheld	adj.（形容词）	掌上型；手持型
navigation	n.（名词）	导航
ongoing	adj.（形容词）	即将
storage	n.（名词）	存储；仓库；贮藏所
terminals	n.（名词）	终端；终端机
wireless	adj.（形容词）	无线的；无线电的

Phrases

charged with	承担，负责
core for	核心
Foundation Profile	基础配置文件
Game Profile	游戏配置文件
industrial-strength	工业级强度；工业级
Java Community Process Program	Java 社团处理程序
memory availability	内存可用性
Mobile Information Device Profile	移动信息装备配置文件
PDA Profile	个人助理配置文件
Personal Basis Profile	个人基本信息配置文件
Personal Profile	个人配置文件
point-of-sale terminals	（计算机）销售点终端机
RMI Profile	RMI 配置文件
set-top boxes	机顶盒
stripped-down	简装的
used with	使用

批注

注1　J2ME 为主语；谓语：must service；including screen phones, digital set-top boxes used for cable television, cell phones, and personal digital assistants. 为分词结构；used for cable television, cell phones, and personal digital assistants 是后置定语，修饰 boxes。

注2　短语：One；the other：是对 two configurations 的补充说明。

注3　主语：CLDC devices；谓语：use；宾语：(KVM) implementation，which 引导一个非限制性定语从句。

注4　主语：MIDP；两个并列谓语：is 和 contains；That provides 这句为定语从句，先行词是 classes；that runs 这句为定语从句，that 的先行词是 application。

注5　这是一个比较句，better, larger than, found 为分词结构作定语，修饰 resources。

注6　Keep in mind 这个短语作主语；谓语：can access；only if 引导一个条件状语从句；in the JVM 与 in the profile 并列；used by the developer 作定语，修饰 the profile。

Associated Reading

C 常见报错及其含义

Ambiguous operators need parentheses	不明确的运算需要用括号括起来
Ambiguous symbol "xxx"	"xxx"符号具有二义性
Argument list syntax error	参数列表语法错误
Array bounds missing	丢失数组界限符
Array size too large	数组太大
Bad character in parameters	参数中有无效的字符
Bad file name format in include directive	编译预处理中的文件名格式不正确
Bad ifdef directive syntax	编译预处理 ifdef 语法错
Bad undef directive syntax	编译预处理 undef 语法错
Bit field too large	位字段太长
Call of non-function	调用未定义的函数
Call to function with no prototype	调用无原型声明的函数
Cannot modify a const ant and object	不允许修改常量对象
Case outside of switch	case 语句在 switch 外
Case syntax error	case 语法错误
Code has no effect	代码不起作用
Compound statement missing{	复合语句漏掉"{"
Conflicting type modifiers	不明确的类型说明符
Constant expression required	要求常量表达式
Constant out of range in comparison	在比较中常量超出范围
Conversion may lose significant digits	转换时会丢失有效数字位
Could not find file "xxx"	找不到"xxx"文件

Declaration missing ;	声明缺少";"
Default outside of switch	default 出现在 switch 语句之外
Define directive needs an identifier	定义编译预处理需要标识符
Division by zero	用 0 作除数
Do statement must have while	do-while 语句中缺少 while 部分
Enum syntax error	枚举类型语法错误
Enumeration constant syntax error	枚举常数语法错误
Error directive :xxx	错误的编译预处理命令":xxx"
Error writing output file	写输出文件错误
Expression syntax error	表达式语法错误
Extra parameter in call	调用时出现多余参数
File name too long	文件名太长
Function call missing")"	函数调用缺少右括号
Function definition out of place	函数定义位置错误
Function should return a value	函数必须返回一个值
Goto statement missing label	Goto 语句没有标号
Hexadecimal or octal constant too large	十六进制或八进制常数太大
Illegal character "x"	非法字符"x"
Illegal initialization	非法的初始化
Illegal octal digit	非法的八进制数字
Illegal pointer subtraction	非法的指针相减
Illegal structure operation	非法的结构体操作
Illegal use of floating point	非法的浮点运算
Illegal use of pointer	指针使用非法
Improper use of a type def symbol	类型定义符号使用不恰当
In-line assembly not allowed	不允许使用行汇编嵌入
Incompatible storage class	存储类别不相容
Incompatible type conversion	不相容的类型转换
Incorrect number format	错误的数据格式
Incorrect use of default	default 使用不当
Invalid pointer addition	指针相加无效
Logical value required	需要逻辑值
Macro argument syntax error	宏参数语法错误
Macro expansion too long	宏扩展后太长
Mismatched number of parameters in definition	定义中参数个数不匹配
Misplaced break	此处不应出现 break 语句
Misplaced continue	此处不应出现 continue 语句
Misplaced decimal point	此处不应出现小数点

Misplaced elif directive	elif 指令位置错
Misplaced else	此处不应出现 else
Misplaced else directive	此处不应出现编译预处理 else
Misplaced endif directive	此处不应出现编译预处理 endif
Must be addressable	必须是可以编址的
Must take address of memory location	必须存储定位的地址
No declaration for function "xxx"	没有函数"xxx"的说明
No stack	缺少堆栈
No type information	没有类型信息
Not a valid expression format type	不合法的表达式格式
Not an allowed type	不允许使用的类型
Numeric constant too large	常量数值太大
Out of memory	超出内存范围
Parameter "xxx" is never used	参数"xxx"没有用到
Pointer required on left side of –>	符号–>的左边必须是指针
Possible use of "xxx" before definition	在定义之前使用了"xxx"
Possibly incorrect assignment	赋值可能不正确
Re-declaration of "xxx"	重复定义了"xxx"
Redefinition of "xxx" is not identical	没有用到参数"xxx"
Register allocation failure	寄存器分配失败
Repeat count needs an l value	重复计数需要逻辑值1
Size of structure or array not known	结构体或数组大小不确定
Statement missing ;	语句后缺少";"
Structure or union syntax error	结构体或联合体语法错误
Structure size too large	结构体尺寸太大
Sub scripting missing]	下标缺少"]"
Superfluous & with function or array	函数或数组中有多余的"&"
Suspicious pointer conversion	可疑的指针转换
Symbol limit exceeded	符号超限
Too few parameters in call	函数调用时的实参少于函数定义的形参
Too many default cases	default 太多(switch 语句中一个)
Too many error or warning messages	错误或警告信息太多
Too many type in declaration	声明中的类型太多
Too much global data defined in file	文件中的全局数据太多
Two consecutive dots	两个连续的句点
Type mismatch in parameter "xxx"	参数"xxx"类型不匹配
Type mismatch in re-declaration of "xxx"	"xxx"重定义的类型不匹配
Unable to create output file "xxx"	无法建立输出文件"xxx"

Unable to open include file "xxx"	无法打开被包含的文件"xxx"
Unable to open input file "xxx"	无法打开输入文件"xxx"
Undefined label "xxx"	没有定义的标号"xxx"
Undefined structure "xxx"	没有定义的结构体"xxx"
Undefined symbol "xxx"	没有定义的符号"xxx"
Unexpected end of file in comment started on line xxx	源文件在 xxx 行开始的注释中意外结束
Unexpected end of file in conditional started on line xxx	源文件在 xxx 行开始的条件语句中意外结束
Unknown assemble instruction	未知的汇编结构
Unknown option	未知的选项
Unreachable code	执行不到的代码
User break	用户强行中断
Void functions may not return a value	void 类型的函数不应有返回值
Wrong number of arguments	调用函数的参数个数错
"xxx" not an argument	"xxx"不是参数
"xxx" not part of structure	"xxx"不是结构体的一部分

Unit 3   Discrete Mathematics

Text A   About Discrete Mathematics

1. Introduction to discrete mathematics

Discrete❶ Mathematics is **the general term**❷ for several branches of mathematics, which is based on the study of mathematical structures that are fundamentally discrete rather than continuous. **In contrast to**❸ real numbers that have the **property**❹ of varying "smoothly", the objects studied in discrete mathematics-such as integers, graphs, and statements in logic-do not vary smoothly in this way, but have **distinct**❺, separated values.注1 Discrete mathematics therefore **excludes**❻ topics in "**continuous mathematics**❼" such as **calculus**❽ and analysis. Discrete objects can often be **enumerated**❾ by integers. More formally, discrete mathematics has been characterized as the branch of mathematics dealing with **countable sets**❿ (sets that have the same **cardinality**⓫ as **subsets**⓬ of the integers, including **rational**⓭ numbers but not real numbers). However, there is no exact, universally agreed, definition of the term "discrete mathematics." Indeed, discrete mathematics is described **less** by what is included **than**⓮ by what is excluded: continuously varying quantities and related **notions**⓯.

Research in discrete mathematics increased in the latter half of the twentieth century partly due to the development of digital computers which operate in discrete steps and store

---

❶ 离散的
❷ 一般术语
❸ 相反地
❹ 性质,性能
❺ 清楚的,有区别的
❻ 排除,排斥
❼ 连续数学
❽ 微积分学
❾ 枚举
❿ 可数集
⓫ 基数
⓬ 子集
⓭ 有理数
⓮ 少于
⓯ 概念

data in discrete **bits**⑯. Concepts and **notations**⑰ from discrete mathematics are useful in studying and describing **objects**⑱ and problems **in branches of**⑲ computer science, such as computer **algorithms**⑳, programming languages, **cryptography**㉑, automated theorem proving, and software development. **Conversely**㉒, computer **implementations**㉓ are significant.

2. Topics in discrete mathematics

Theoretical computer science includes areas of discrete mathematics relevant to computing. It draws heavily on graph theory and logic. Included within theoretical computer science is the study of algorithms for computing mathematical results. **Computability**㉔ studies what can be computed in principle, and has close ties to logic, while **complexity studies**㉕ the time taken by computations. Automata theory and formal language theory are closely related to computability. Computational geometry applies algorithms to **geometrical**㉖ problems, while computer image analysis applies them to representations of images. Theoretical computer science also includes the study of various continuous computational topics.

Logic is the study of the principles of valid reasoning and **inference**㉗, **as well as**㉘ of **consistency soundness**㉙, and **completeness**㉚. For example, in most systems of logic (but not in **intuitionistic logic**㉛) Peirce's law (((P→Q)→P)→P) is a **theorem**㉜. For classical logic, it can be easily **verified**㉝ with a truth table. The study of mathematical proof is particularly important in logic, and has applications to **automated**㉞ theorem proving and

---

⑯ 位数
⑰ 符号
⑱ 对象,物体
⑲ 分支
⑳ 算法
㉑ 密码学
㉒ 相反地
㉓ 实施
㉔ 可计算性
㉕ 复杂研究
㉖ 几何的
㉗ 推断,推理
㉘ 也
㉙ 一致性,可靠性
㉚ 完整性
㉛ 直觉主义逻辑
㉜ 定理,原理
㉝ 验证
㉞ 使自动化

formal **verification**❸❺ of software.

Logical formulas are discrete structures, as are proofs, which form finite trees or, more generally, **directed acyclic graphstructures**❸❻ (with each inference step combining one or more **premise**❸❼ branches to give a single conclusion). The truth values of logical formulas usually form a finite set, generally restricted to two values: true and false, but logic can also be continuous-valued, e. g., fuzzy logic. Concepts such as infinite proof trees or infinite **derivation**❸❽ trees have also been studied, e. g. infinitary logic.

Set theory is the branch of mathematics that studies sets, which are collections of objects, such as {blue, white, red} or the (infinite) set of all prime numbers. **Partially**❸❾ ordered sets and sets with other relations have applications in several areas.

In discrete mathematics, countable sets (including finite sets) are the main focus. The beginning of set theory as a branch of mathematics is usually marked by Georg Cantor's work distinguishing between different kinds of infinite set, motivated by the study of **trigonometric series**❹⓪, and further development of the theory of infinite sets is outside **the scope of**❹❶ discrete mathematics. ^注2 Indeed, **contemporary work**❹❷ in descriptive set theory makes extensive use of traditional continuous mathematics.

Graph theory, the study of graphs and networks, is often considered part of combinatory, but has grown large enough and **distinct**❹❸ enough, with its own kind of problems, to **be regarded as**❹❹ a subject in its own right which **in all areas of**❹❺ math and science have extensive application.

Graphs are one of the prime objects of study in Discrete Mathematics. They are among the most **ubiquitous**❹❻ models of both natural and human-made structures. They can model many types of relations and process dynamics in physical, biological and social systems. In computer science, they represent networks of communication, data organization, computational devices, the flow of computation, etc. In Mathematics, they are useful in geometry and certain parts of **topology**❹❼, e. g. **Knot Theory**❹❽. Algebraic graph theory

---

❸❺ 验证
❸❻ 有向无环图结构
❸❼ 前提
❸❽ 引出,来历
❸❾ 部分地
❹⓪ 三角级数
❹❶ 范围
❹❷ 当代作品
❹❸ 区分
❹❹ 看作
❹❺ 在所有的领域
❹❻ 普遍存在的
❹❼ 拓扑学
❹❽ 结点理论

has close links with group theory. There are also continuous graphs, however for the most part research in graph theory falls within the **domain**⑭ of discrete mathematics.

Operations research provides techniques for solving practical problems in business and other fields—problems such as allocating resources to maximize profit, or scheduling project activities to minimize risk. Operations research techniques include linear programming and other areas of **optimization**⑮, queuing theory, scheduling theory, network theory. Operations research also includes continuous topics such as **continuous-time Markov process**⑯, **continuous-time martingales**⑰, **process optimization**, **and continuous and hybrid**⑱ control theory.

Although topology is the field of mathematics that **formalize**⑲ and **generalizes**⑳ the **intuitive**㉑ notion of "**continuous deformation**㉒" of objects, it gives rise to many discrete topics; this can be attributed in part to the focus on topological invariants, which themselves usually take discrete values. See **combinatorial topology**㉓, **topological graph theory**㉔, **topological combinatorics**㉕, **computational topology**㉖, **discrete topological space**㉗, **finite topological space**㉘.

Words

algorithm	*n.*（名词）	算法;算法式(algorithm 的复数)
automate	*adj.*（形容词）	自动化的;机械化的
bit	*n.*（名词）	比特,位
calculu	*n.*（名词）	结石;微积分学
cardinality	*n.*（名词）	基数;集的势
computability	*n.*（名词）	可计算性
conversely	*adv.*（副词）	相反地

---

⑭ 领域
⑮ 优化
⑯ 连续时间马尔可夫过程
⑰ 连续时间鞅
⑱ 过程优化及连续混合控制理论
⑲ 形式化
⑳ 一般化
㉑ 直觉
㉒ 连续变形
㉓ 组合拓扑
㉔ 拓扑图论
㉕ 拓扑组合
㉖ 计算拓扑
㉗ 离散空间
㉘ 有限拓扑空间

cryptography	*n.*（名词）	密码学；密码使用法
derivation	*n.*（名词）	引出；来历；词源
discrete	*adj.*（形容词）	离散的，不连续的
distinct	*adj.*（形容词）	明显的；独特的；清楚的；有区别的
domain	*n.*（名词）	领域；域名；产业；地产
enumerate	*vt.*（名词）	枚举，列举
exclude	*vt. & vi.*（动词）	排除；排斥；拒绝接纳；逐出
formalize	*vt. & vi.*（动词）	使形式化；使正式；拘泥礼仪
generalize	*vt. & vi.*（动词）	概括；推广；使……一般化
geometrical	*adj.*（形容词）	几何的，几何学的
hybrid	*n.*（名词）	混合
implementation	*n.*（名词）	实施
inference	*n.*（名词）	推理；推论；推断
intuitive	*adj.*（形容词）	直觉的；凭直觉获知的
notation	*n.*（名词）	符号；记法
notion	*n.*（名词）	观念；小商品
object	*n.*（名词）	物体；对象
optimization	*n.*（名词）	最佳化，最优化
partially	*adv.*（副词）	部分地；偏袒地
premise	*vt. & vi.*（动词）	引出，预先提出；作为……的前提
property	*n.*（名词）	性质，性能；财产；所有权
rational	*adj.*（形容词）	合理的；理性的
subsets	*n.*（名词）	子集合（subset 的复数）
theorem	*n.*（名词）	定理原理
topology	*n.*（名词）	拓扑学；地质学；局部解剖学
ubiquitous	*adj.*（形容词）	普遍存在的；无所不在的
verification	*n.*（名词）	确认，查证；核实
verified	*adj.*（形容词）	已查清的，已证实的

Phrases

as well as	也；和……一样；不但……而且
be regarded as	被认为是；被当作是
combinatorial topology	组合拓扑学；组合拓扑
complexity studies	复杂性研究
computational topology	计算拓扑
consistency soundness	一致性，可靠性
contemporary work	当代作品

continuous deformation	连续形变
continuous mathematics	连续数学
continuous-time Markov process	连续时间的马尔可夫过程
continuous-time martingales	连续时间鞅
countable sets	可数集
directed acyclic graphstructures	有向无环图结构
discrete topological space	离散拓扑空间
distinct enough	截然不同
finite topological space	有限拓扑空间
in all areas of	在各方面的
in branches of	分支的
in contrast to	相比之下
intuitionistic logic	直觉主义逻辑
knot theory	结点理论；[数]纽结理论
the general term for	一般的术语
the scope of	范围
topological combinatorics	拓扑组合
topological graph theory	图论拓扑
trigonometric series	三角级数

Exercises

【Ex1】 **Answer the questions according to the text：**

(1) What is Discrete Mathematics?
(2) Why did Discrete Mathematics develop so fast in the twentieth century?
(3) How many topics are there in this chapter, and what are they?
(4) What are the logic formulas?
(5) What does Operations research involve?

【Ex2】 **Translate into Chinese：**

(1) Graph theory is an old subject with modern applications.
(2) Relation between elements of sets is represented using the structure called a relation.
(3) Much of discrete mathematics is developed of discrete structures, which are used to represent discrete objects.
(4) Discrete mathematics is the gateway to more advanced courses in all parts of the mathematical sciences.

(5) The computer chip is primarily responsible for executing instructions.

(6) Tape must be read or written sequentially, not randomly.

(7) Deselect the text by clicking anywhere outside of the selection on the page or pressing an arrow key on the keyboard.

(8) Faster than many types of parallel port, a single USB port is capable of chaining many devices without the need of a terminator.

【Ex3】 Choose the best answer:

(1) Very long, complex expressions in program are difficult to write correctly and difficult to _____.

    A. defend   B. detect   C. default   D. debug

(2) _____ is the study of the principles of valid reasoning and inference, as well as of consistency, soundness, and completeness.

    A. Graph theory   B. Logic
    C. Topology   D. Operation research

(3) The _____ storage area that you can use to copy or move selected text or object among applications.

    A. exponent   B. order   C. temporary   D. superior

(4) Software design is a _____ process. It requires a certain _____ of flair on the part of the designer.

    A. create, amount   B. created, amounted
    C. creating, mount   D. creative, mounted

批 注

注1 In contrast to 短语引导的独立结构;that have the property of varying "smoothly"的先行词为real numbers,属定语从句;the objects 为句子的主语;studied 是定语,do not vary 是谓语;do not vary smoothly in this way, but have distinct, separated values. 补充说明。

注2 The beginning of set theory 为主语;as a branch of mathematics 为方式状语;is marked, motivated 为句子谓语。

Text B　Tree

A **connected graph**❶ shown in Figure 3-1, that contains no simple **circuits**❷ is called a tree. Trees were used as long ago as 1857, when the English mathematician Arthur Cayley

---

❶ 连通图
❷ 电路,回路

used them to count certain types of chemical **compounds** ❸. Since that time, trees have been employed to solve problems **in a wide variety of**❹ **disciplines**❺.

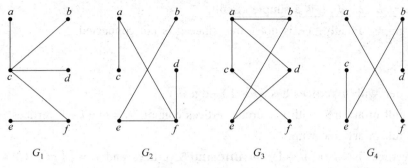

Figure 3-1   Graphs

Tree are particularly useful in computer science, for instance, trees **are employed to**❻ construct efficient algorithms for locating **items**❼ in a list. They are used to construct network with the least expensive set of telephone lines storing and **transmitting**❽ data. 注1 Trees can model procedures that **are carried out**❾ using a sequence of decisions. This makes trees valuable **in the study**❿ of sorting algorithms.

1. Definitions

A tree is an undirected simple graph G that satisfies any of the following **equivalent**⓫ conditions注2:

- G is connected and has no cycles.
- G has no cycles, and a simple cycle is formed if any edge is added to G.
- G is connected, and it is not connected anymore if any edge is removed from G.
- G is connected and the 3-vertex complete graph K_3 is not a minor of G.注3
- Any two **vertices**⓬ in G can be connected by a unique simple path.

If G has finitely many vertices, say n of them, then the above statements are also equivalent to any of the following conditions:

- G is connected and has $(n-1)$ edge.
- G has no simple cycles and has $(n-1)$ edge.

- ❸ 混合物
- ❹ 以更广泛的
- ❺ 学科,分支
- ❻ 用于
- ❼ 条款,项目
- ❽ 传播;发射
- ❾ 执行
- ❿ 研究
- ⓫ 等价的
- ⓬ 结点

Which of the graphs shown in Figure 3-1 are trees?

G_1 and G_2 are trees, since both are connected graphs with no simple circuits; G_3 is not a tree because $e.b.a.d.e$ is a simple circuit. 注4

In this graph, Finally, G_4 is not a tree since it is not connected.

2. Facts

(1) A tree with n vertices has $(n-1)$ edges.

(2) A full m-ary tree with i internal vertices contains $n = m*i+1$ vertices.

(3) A full m-ary tree with.

(i) n vertices has $i = (n-1)/m$ **internal**⑬ vertices and $L = [(m-1)*n+1]/m$ leaves.

(ii) i internal vertices has $n = m*i+1$ vertices and $L = (m-1)*i+1$ leaves.

(iii) L leaves has $n = (m*L-1)/(m-1)$ vertices and $i = (L-1)/(m-1)$ internal vertices.

3. Tree spices

(1) M-ary tree is a tree with the property that every internal **vertex**⑭ has no more than m children.

(2) Binary tree is an m-ary tree with m = 2 (each child may be designated as a left or a right child of its parent).

(3) Ordered tree is a tree in which the children of each internal vertex are **linearly**⑮ ordered. 注5

(4) Balance tree is a tree in which every vertex is at level h or h-1. where h is the height of the tree. 注6

(5) Binary search tree is a binary tree in which the vertices are labeled with items **so that**⑯ a label of a vertex is greater than the labels of all vertices in the left subtree of this vertex is an less than the label of all vertices in the right sub tree of this vertex.

(6) Decision tree is a rooted tree where each vertex represents a possible outcome of a decision and the leaves represent the possible solutions.

(7) **Spanning tree**⑰: a tree containing all vertices of a graph.

(8) **Minimum spanning tree**⑱: a spanning tree with smallest possible sum of weights of its edges. 注7

---

⑬ 内部的
⑭ 结点
⑮ 呈直线地
⑯ 致使
⑰ 生成树
⑱ 最小生成树

4. Tree traversal algorithms

Procedures for systematically visiting every vertex of an ordered rooted tree are called **traversal algorithms**[19]. We will describe three of the most commonly used such algorithms preorder traversal, inorder traversal, and postorder traversal.

(1) Preorder traversal

Preorder traversal[20] is a listing of the vertices of an ordered rooted tree defined recursively by specifying that the root is listed, Followed by the first subtree, followed by the other sub trees in the order they occur from left to right.

Let T be an ordered rooted tree with root r. if T consists only of r, then r is the preorder traversal of T. Otherwise, suppose that $T_1, T_2, \cdots, T_n$ are the **sub trees**[21] at r from left to right in T. The preorder traversal begins by visiting r. It continues by traversing T_1 in preorder, then T_2 in preorder, and so on, until T_n is traversed in preorder.

(2) Inorder traversal

Inorder traversal[22] is **a listing of**[23] the vertices of an ordered rooted tree **defined recursively by**[24] specifying that the first sub tree is listed **followed by**[25] the root, followed by the other sub trees in the order they occur from left to right.

Let T be an ordered rooted tree with root r. if T consists only of r, then r is the inorder traversal of T. Otherwise, suppose that $T_1, T_2, \cdots, T_n$. are the subtrees at r from left to right in T. The inorder traversal begins by traversing T_1 in inorder. Then visiting r, It continues by traversing T_2 in inorder, then T_3 in inorder, $\cdots$, And finally T_n in inorder.

(3) Postorder traversal

Postorder traversal[26] is a listing of the vertices of an ordered rooted tree defined recursively by specifying that the sub trees are listed in the order they occur from left to right, followed by the root. 注8

Let T be an ordered rooted tree with root r if T consists only of r, then r is the post order traversal of T. Otherwise, suppose that $T_1, T_2, \cdots, T_n$. are the subtrees at r from left to right. The postorder traversal begins by traversing T_1 in postorder, then T_2 in postorder, $\cdots$, Then T_n. in postorder, and ends by visiting r.

---

[19] 遍历算法
[20] 前序遍历
[21] 子树
[22] 中序遍历
[23] 列表
[24] 被定义为
[25] 紧跟着
[26] 后序遍历

Words

circuit	n.（名词）	电路；环路；巡回
compound	vt. & vi.（动词）	合成；混合；和解妥协；掺和
connected graph	n.（名词）	连接图，[数]连通图
discipline	n.（名词）	纪律
equivalent	adj.（形容词）	等价的，相等的；同意义的
inorder traversal	n.（名词）	中序遍历
internal	adj.（形容词）	内部的；内在的；国内的
item	n.（名词）	项目；名目；所有物品
preorder traversal	n.（名词）	先序遍历
transmit	vt. & vi.（动词）	传递，发射
vertex	n.（名词）	顶点；[昆]头顶；[天]天顶
vertice	n.（名词）	制高点；天顶；头顶

Phrases

be carried out	进行
be employed to	用来
defined by	定义为
followed by	紧随其后
in a wide variety of	各种各样的
in the study	在这项研究中

批 注

注1　storing and transmitting data 用于修饰 network，作定语。此句话的含义是：他们用花销代价最少的电话线构建网络，用它来存储、传输数据。

注2　此句话的含义是：一个树是一个无向简单图 G，它满足以下任何等价条件。

注3　G 是连通的，3 个顶点构成了图 K_3，它不是一个 G 图的镜像图。

注4　图 G_1 和图 G_2 是树，因为它们都是有向图，没有回路的有向图，而图 G_3 不是树，因为 ebade 是一个简单的回路。

注5　主语：Ordered tree；系动词：is；in which 引导一个定语从句；the children 为定从的主语；are ordered 是定从的谓语。这句话的含义是：有序树是它的内部子顶点都是线性地排序的树。

注6　主语：Balance tree；系动词：is；in which 引导一个定语从句；every vertex 为从句的主语。这句话的含义是：平衡树是它的每个顶点都在 h 或 $h-1$ 层，而 h 是该树的高度。

注7　最小生成树：它的边权总数为最小的树。

注8　主语：Postorder traversal；谓语：is；表语：a listing of the vertices of an ordered rooted tree；by 引导状语主语。

Associated Reading

Topics in Discrete Mathematics

Discrete mathematics is the study of mathematical structures that are fundamentally discrete rather than continuous. The main topics in discrete mathematics have as the following:

(a) Theoretical computer science

(b) Information theory

(c) Logic

(d) Set theory

(e) Combinatorics

(f) Graph theory

(g) Probability

(h) Number theory

(i) Algebra

(j) Calculus of finite differences, discrete calculus or discrete analysis

(k) Geometry

(l) Topology

(m) Operations research

(n) Game theory, decision theory, utility theory, social choice theory

(o) Discretization

(p) Discrete analogues of continuous mathematics

(q) Hybrid discrete and continuous mathematics

相关离散数学主题：

(a) 理论计算机科学

(b) 信息论

(c) 逻辑学

(d) 集合论

(e) 组合数学

(f) 图论

(g) 概率论

(h) 数论

(i) 代数

(j) 差分演算,离散模型符号验证或者离散分析

(k) 几何学

(l) 拓扑学

(m) 运筹学
(n) 博弈论、决策论、效用理论、社会选择理论
(o) 离散化
(p) 连续数学的离散近似
(q) 离散和连续混合数学

Unit 4　Software Engineering

Text A　Software Processes

　　A software process is a set of activities that leads to the production of a software product. These activities may involve the development of software from **scratch**❶ in a standard programming language like Java or C. Increasingly, however, new software is developed by extending and modifying existing systems and by **configuring**❷ and **integrating**❸ off-the-shelf software or system components.注1

　　Software processes are complex and, like all **intellectual**❹ and creative processes, rely on people making decisions and judgments. **Because of**❺ the need for judgment and creativity, attempts to **automate**❻ software processes **have met with**❼ limited success. Computer-aided software engineering (CASE) tools can support some process activities. However, there is no possibility, **at least**❽ in the next few years, of more extensive automation where software **takes over**❾ creative design from the engineers involved in the software process.注2

　　One reason the effectiveness of CASE tools is limited is because of the **immense**❿ diversity of software processes. There is no ideal process, and many organizations have developed their own approach to software development. Processes have evolved to exploit the capabilities of the people in an organization and the specific characteristics of the systems that are being developed. For some systems, such as critical systems, a very structured development process is required. For business systems, with rapidly changing requirements, a flexible, agile process is likely to be more effective.

　　Although there are many software processes, some **fundamental**⓫ activities are common to all software processes:

---

❶ 从零开始
❷ 配置
❸ 合并;整合
❹ 智力的
❺ 由于
❻ 使自动化
❼ 实现;遇见
❽ 至少
❾ 接管,负责
❿ 极大的
⓫ 基本的

(1) **Software specification** the functionality of the software and **constraints**❷ on its operation must be defined.

(2) **Software design and implementation** the software to meet the specification must be produced.

(3) **Software validation**❸ the software must be validated to ensure that it does what the customer wants.

(4) **Software evolution** the software must evolve to meet changing customer needs.

Although there is no 'ideal' software process, there is **scope**❹ for improving the software process in many organizations. Processes may include outdated techniques or may not take advantage of the best practice in industrial software engineering. Indeed, many organizations still do not **take advantage of**❺ software engineering methods in their software development.

Software processes can be improved by process standardization where the **diversity**❻ in software processes across an organization is reduced. This leads to improved communication and a reduction in training time, and makes automated process support more economical. Standardization is also an important first step in introducing new software engineering methods and techniques and good software engineering practice.

Software process models

A software process model is an abstract representation of a software process. Each process model represents a process from a particular **perspective**❼, and thus provides only **partial**❽ information about that process.注3 In this section, I introduce a number of very general process models (sometimes called process **paradigms**❾) and present these from an **architectural perspective**❿. That is, we see the framework of the process but not the details of specific activities.

These **generic**㉑ models are not **definitive**㉒ descriptions of software processes. Rather, they are abstractions of the process that can be used to explain different approaches to

---

❷ 约束
❸ 有效性
❹ 范围
❺ 充分利用
❻ 差异
❼ 观点;客观判断力
❽ 部分的
❾ 范例
❿ 体系结构角度
㉑ 一般的;普通的
㉒ 最佳的;最完整可靠的

software development. You can **think of** them **as**㉓ process frameworks that may be extended and adapted to create more specific software engineering processes. 注4

(1) **The waterfall model**㉔ this takes the fundamental process activities of specification, development, validation and evolution and represents them as separate process **phases**㉕ such as requirements **specification**㉖, software design, implementation, testing and so on. 注5

(2) **Evolutionary development**㉗ this approach **interleaves**㉘ the activities of specification, development and validation. An initial system is rapidly developed from abstract specifications. This is then refined with customer input to produce a system that satisfies the customer's needs.

(3) **Component-based software engineering**㉙ this approach is based on the existence of a significant number of reusable components. The system development process focuses on integrating these components into a system rather than developing them from scratch.

The waterfall model

The first published model of the software development process was **derived**㉚ from more general system engineering processes (Royce, 1970). This is illustrated in Figure 4-1. Because of the **cascade**㉛ from one phase to another, this model is known as the waterfall model or software life cycle. The principal stages of the model map onto fundamental development activities:

i. **Requirements analysis and definition**　the system's services, constraints and goals are established by **consultation**㉜ with system users. They are then defined **in detail**㉝ and **serve as**㉞ a system specification.

ii. **System and software design**　the systems design process **partitions**㉟ the requirements to **either** hardware **or**㊱ software systems. It establishes overall system architecture. Software design involves identifying and describing the fundamental software

---

㉓ 把……认为是
㉔ 瀑布模型
㉕ 阶段
㉖ 规范;说明书
㉗ 演化开发模型
㉘ 交替
㉙ 基于组件的软件工程
㉚ 派生
㉛ 嵌套
㉜ 咨询
㉝ 详尽的
㉞ 作为
㉟ 部分;分割
㊱ 或……或

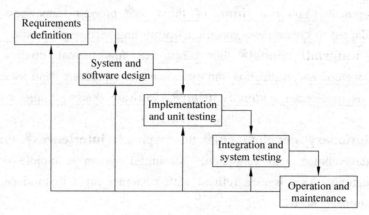

Figure 4-1　The software life cycle

system abstractions and their relationships.

iii. Implementation and unit testing　during this stage, the software design is realized as a set of programs or program units. Unit testing involves **verifying**⑰ that each unit meets its specification.

iv. Integration and system testing　the individual program units or programs are integrated and tested as a complete system to ensure that the software requirements have been met. After testing, the software system is delivered to the customer. 注6

v. Operation and maintenance⑱　normally (although not necessarily) this is **the longest life-cycle phase**⑲. The system is installed and put into practical use. Maintenance involves correcting errors which were not discovered in earlier stages of the life cycle, improving the implementation of system units and enhancing the system's services as new requirements are discovered. 注7

In principle, the result of each phase is one or more documents that are approved ('Signed off'). The following phase should not start until the previous phase has finished. In practice, these stages **overlap**⑳ and **fleed** information **to**㉑ each other. 注8 During design, problems with requirements are identified; During coding design problems are found and **so on**㉒. The software process is not a simple linear model but involves a sequence of **iterations**㉓ of the development activities.

Because of the costs of producing and approving documents, iterations are costly and

---

⑰　核实；核准
⑱　操作与维护
⑲　最长的生存阶段
⑳　部分重叠
㉑　反馈，传递
㉒　等等
㉓　迭代

involve significant **rework**⓮. Therefore, after a small number of iterations, it is normal to freeze parts of the development, such as the specification, and to continue with the later development stages.^注9 Problems are left for later resolution, ignored or programmed around. This premature freezing of requirements may mean that the system won't do what the user wants. It may also lead to badly structured systems as design problems are **circumvented by**⓯ implementation **tricks**⓰.

During the final life-cycle phase (operation and maintenance), the software is put into use. Errors and **omissions**⓱ in the original software requirements are discovered. Program and design errors emerge and the need for new functionality is identified. The system must therefore evolve to remain useful.^注10 Making these changes (software maintenance) may involve repeating previous process stages.

The advantages of the waterfall model are that documentation is produced at each phase and that it fits with other engineering process models. Its major problem is its **inflexible**⓲ partitioning of the project into distinct stages. **Commitments**⓳ must be made at an early stage in the process, which makes it difficult to respond to changing customer requirements.

Therefore, the waterfall model should only be used when the requirements are well understood and unlikely to change **radically**⓴ during system development.

However, the waterfall model reflects the type of process model used in other engineering projects. Consequently, software processes based on this approach are still used for software development, particularly when the software project is part of a larger systems engineering project.

Words

automate	vt. & vi.（动词）	自动化,自动操作
cascade	n.（名词）	嵌套
circumvented	vt. & vi.（动词）	包围;陷害;绕行
commitment	n.（名词）	承诺,保证;委托;承担义务;献身
configure	vt. & vi.（动词）	配置;使成形
constraint	n.（名词）	约束;限制;约束条件
consultation	n.（名词）	咨询;磋商;[临床]会诊;讨论会

⓮ 重做
⓯ 由……包围
⓰ 技巧;诡计
⓱ 省略
⓲ 死板的;硬的
⓳ 承诺;保证
⓴ 完全地;彻底地

definitive	*adj.*（形容词）	决定性的；最后的；限定的
derived	*vt. & vi.*（动词）	得到；推断
fundamental	*adj.*（形容词）	基本的，根本的
generic	*adj.*（形容词）	类的；一般的；属的；非商标的
immense	*adj.*（形容词）	巨大的，广大的；无边无际的
inflexible	*adj.*（形容词）	顽固的；不可弯曲的
integrating	*vt. & vi.*（动词）	整合；积分；集成化
intellectual	*adj.*（形容词）	智力的；聪明的；理智的
interleave	*vt.*（动词）	交替
iteration	*n.*（名词）	迭代次数；反复
omission	*n.*（名词）	疏忽，遗漏；省略；冗长
overlap	*vt. & vi.*（动词）	重叠；重复
paradigm	*n.*（名词）	范例，模范
partial	*adj.*（形容词）	局部的；偏爱的；不公平的
partition	*vt. & vi.*（动词）	分开
perspective	*n.*（名词）	观点，角度
phase	*n.*（名词）	阶段，时期
radically	*adv.*（副词）	根本上；彻底地；以激进的方式
scope	*n.*（名词）	范围
scratch	*n.*（名词）	草稿
specification	*n.*（名词）	规格；说明书；详述
validation	*n.*（名词）	确认；批准；生效
verifying	*vt. & vi.*（动词）	验证；核查

Phrases

architectural perspective	体系结构角度
at least	至少
be delivered to	送到
because of	因为；由于
component-based software engineering	基于组件的软件工程
evolutionary development	演进式开发
fleed…to	反馈，传递
in detail	详细地
in earlier stages of the life cycle	在生命周期的前期
take advantage of	利用
takes over	接管
the longest life-cycle phase	最长的生命周期阶段

| the waterfall model | 瀑布模型 |
| think of | 记起,想起;考虑;想象;关心 |

Exercises

【Ex1】 Answer the questions according to the text：

(1) What is the software process?

(2) What are the fundamental activities common to all software processes?

(3) How to improve the software process?

(4) Why do we need to freeze parts of the development after a small number of iterations?

(5) Why must the system evolve to remain useful?

【Ex2】 Translate into Chinese：

(1) A software process is a set of activities that leads to the production of a software product.

(2) Because of the need for judgement and creativity, attempts to automate software processes have met with limited success.

(3) One reason the effectiveness of CASE tools is limited is because of the immense diversity of software processes.

(4) For business systems, with rapidly changing requirements, a flexible, agile process is likely to be more effective.

(5) A software process model is an abstract representation of a software process.

(6) The system development process focuses on integrating these components into a system rather than developing them from scratch.

(7) During this stage, the software design is realized as a set of programs or program units.

(8) The advantages of the waterfall model are that documentation is produced at each phase and that it fits with other engineering process models.

【Ex3】 Choose the best answer

(1) _____ means "Any HTML document a HTTP Server".

    A. Web server   B. Web page

    C. Web browser   D. Web site

(2) The term "_____ program" means a program written in high-level language.

    A. compiler   B. executable   C. source   D. object

(3) Very long complex expressions in a program are difficult to write correctly and

difficult to _____.

    A. defend   B. detect   C. default   D. debug

（4）In C language, functions are important because they provide a way to _____ code so that a large complex program can be written by combining many smaller parts.

    A. modify   B. modularize   C. block   D. board

（5）The standard _____ in C language contain many useful functions for input and output, string handing, mathematical computations, and system programming tasks.

    A. databases   B. files   C. libraries   D. subroutines

批 注

注1　主语：new software；谓语：is developed；状语：by extending and modifying existing systems and by configuring and integrating off-the-shelf software or system components。这句话的含义是：逐渐地，现在越来越多的软件是通过在旧软件基础上修改或通过配置和集成现成软件或系统组件而形成的。

注2　这句话的含义是：但是，更广泛的、自动化的可以替代工程师在软件过程中的有创造力的设计至少在今后几年内不可能出现；at least in the next few years 作为插入语；possibility of：……的可能。

注3　主语：Each process model；谓语：represents，provides。这句话的含义是：每个过程模型从一个特定的角度表现一个过程，只提供过程的某一方面的信息。

注4　that 引导一个定语从句；先行词是 process frameworks。这句话的含义是：你可以将这些模型看作过程处理框架，可以扩展并修改它们创建一个特殊的软件工程处理过程。

注5　这句话的含义是：这个模型采用一些基本的过程活动，即描述、开发、有效性验证和进化，代表单独的过程阶段（如需求描述、软件设计、实现和测试等阶段）表现这些活动。

注6　这句话的含义是：集成单个的程序单元或程序，并对系统整体进行测试，以满足需求。测试之后，将软件系统交付给客户使用。

注7　Maintenance 为主语；involves 为谓语；correcting，improving，enhancing 为并列结构。

注8　这句话的含义是：在实践中，这些阶段相互重叠，彼此间有信息交换。

注9　这句话的含义是：因此，经过少量的反复之后，要冻结部分开发过程，如描述部分，继续进行后面的开发阶段。

注10　这句话的含义是：因此，系统必须进化，以保持实用性。

Text B　Introducing the UML

    The **Unified Modeling Language**❶（UML）is a standard language for writing software **blueprint**❷. The UML may be used to visualize, **specify**❸, construct, and document the **artifacts**❹ of a **software- intensive system**❺.

---

❶ 统一建模语言
❷ 蓝图
❸ 详述
❹ 人工制品
❺ 软件密集型系统

The UML is appropriate for modeling systems ranging from **enterprise information systems**❻ to distributed **Web-based applications**❼ and even to hard **real time embedded systems**❽. It is a very expressive language, addressing all the views needed to develop and then **deploy**❾ such systems. Even though it is expressive, the UML is not difficult to understand and to use. Learning to apply the UML effectively starts with forming a **conceptual**❿ model of the language, which requires learning three major elements: the UML's basic building blocks, the rules that **dictate**⓫ how these building blocks may be put together, and some common **mechanisms**⓬ that apply throughout the language. 注1

The UML is only a language, so it is just one part of a software development method. The UML is process independent, although **optimally**⓭ it should be used in a process that is case driven, architecture-centric, **iterative**⓮, and **incremental**⓯.

1. An overview of the UML

The UML is a language for
(1) Visualizing　　(2) Specifying　　(3) Constructing　　(4) Documenting

2. The UML is a language

A language provides a vocabulary and the rules for combining words in that vocabulary for the purpose of communication. A modeling language is a language whose vocabulary and rules focus on the conceptual and physical **representation**⓰ of a system. A modeling language such as the UML is thus a standard language for software blueprints.

Modeling **yields**⓱ an understanding of a system. No one model is ever sufficient. Rather, you often need **multiple**⓲ models that are connected to one another to understand anything but the most **trivial**⓳ system. For software-intensive systems, this requires a language that addresses the different views of a system's architecture as it evolves throughout the software development life cycle.

---

❻ 企业信息系统
❼ 网站应用系统
❽ 实时嵌入式系统
❾ 使展开,应用
❿ 概念的
⓫ 支配,主宰
⓬ 机制
⓭ 最佳
⓮ 迭代
⓯ 增量
⓰ 代表
⓱ 产生
⓲ 倍数
⓳ 琐碎;细小

The vocabulary and rules of a language such as the UML tell you how to create and read well-formed models, but they don't tell you what models you should create and when you should create them. That's the role of the software development process. A well-defined process will guide you in deciding what artifacts to produce, what activities and what workers to use to create them and manage them, and how to use those artifacts to measure and control the project as a whole.^{注2}

3. The UML is a language for visualizing

For many programmers, the distance between thinking of an **implementation**[20] and then pounding it out in code is close to zero. You think it, you code it. In fact, some things are best cast directly in code. Text is a wonderfully **minimal**[21] and direct way to write expressions and **algorithms**[22].

In such cases, the programmer is still doing some modeling, **albeit**[23] entirely mentally. He or she may even sketch out a few ideas on **a white board**[24] or on **a napkin**[25]. However, there are several problems with this. First, communicating those conceptual models to others is **error-prone**[26] unless everyone involved speaks the same language.^{注3} Typically, projects and organizations develop their own language, and it is difficult to understand what's going on if you are an outsider or new to the group. Second, there are some things about a software system you can't understand unless you build models that **transcend**[27] the textual programming language. For example, the meaning of a class **hierarchy**[28] can be inferred, but not directly grasped, by staring at the code for all the classes in the hierarchy. Similarly, the physical distribution and possible **migration**[29] of the objects in a Web-based system can be inferred, but not directly grasped, by studying the system's code.^{注4} Third, if the developer who cut the code never wrote down the models that are in his or her head, that information would be lost forever or, at best, only partially re-creatable from the implementation once that developer moved on.

Writing models in the UML addresses the third issue: an explicit model facilitates communication.

---

  [20] 实施；履行
  [21] 最小的
  [22] 算法
  [23] 尽管
  [24] 白板
  [25] 纸巾
  [26] 易出错
  [27] 超越
  [28] 层次
  [29] 迁移

Some things are best modeled textually; others are best modeled **graphically**③⓪. Indeed, in all interesting systems, there are structures that transcend what can be represented in a programming language. The UML is such a graphical language. This addresses the second problem described earlier.

The UML is more than just a bunch of graphical symbols. 注5 Rather, behind each symbol in the UML notation is a **well-defined**③① **semantics**③②. In this manner, one developer can write a model in the UML, and another developer, or even another tool, can interpret that model **unambiguously**③③. This addresses the first issue described earlier.

4. The UML is a language for specifying

In this context, specifying means building models that are **precise**③④, unambiguous, and complete. In particular, the UML addresses the specification of all the important analysis, design, and implementation decisions that must be made in developing and deploying a software-intensive system.

5. The UML is a language for constructing

The UML is not a visual programming language, but its models can be directly connected to a variety of programming languages. This means that it is possible to map from a model in the UML to a programming language such as Java, C++, or Visual Basic, or even to tables in a relational **database**③⑤ or the **persistent**③⑥ store of an object-oriented database. 注6 Things that are best expressed graphically are done so graphically in the UML, whereas things that are best expressed textually are done so in the programming language.

This **mapping**③⑦ permits forward engineering the generation of code from a UML model into a programming language. The reverse is also possible: You can reconstruct a model from an implementation back into the UML. 注7 Reverse engineering is not magic. Unless you encode that information in the implementation, information is lost when moving forward from models to code. Reverse engineering thus requires tool support with human **intervention**③⑧. Combining these two paths of forward code generate and reverse engineering yields round-trip engineering mean the ability to work in either a graphical or a textual view, while tools keep the two views consistent. 注8

- ③⓪ 图形
- ③① 定义明确的
- ③② 语义
- ③③ 毫不含糊地
- ③④ 精确地
- ③⑤ 数据库
- ③⑥ 持久性
- ③⑦ 映射
- ③⑧ 干预

In addition to this direct mapping, the UML is sufficiently expressive and unambiguous to permit the direct execution of models, the **simulation**[39] of systems, and the **instrumentation**[40] of running systems.

6. The UML is a language for documenting

A healthy software organization produces all sorts of artifacts in addition to raw executable code. These artifacts include (but are not limited to)

      Requirements
      Architecture
      Design
      Source code
      Project plans
      Tests
      Prototypes[41]
      Releases

Depending on the development culture, some of these artifacts are treated more or less formally than others. Such artifacts are not only the deliverables of a project, they are also **critical**[42] in controlling, measuring, and communicating about a system during its development and after its deployment.

The UML addresses the documentation of a system's architecture and all of its details. The UML also provides a language for expressing requirements and for tests. Finally, the UML provides a language for modeling the activities of project planning and release management.

Words

albeit	*n.* (名词)	尽管
artifact	*n.* (名词)	人工制品
blueprint	*n.* (名词)	蓝图
conceptual	*adj.* (形容词)	概念的
critical	*adj.* (形容词)	关键的
database	*n.* (名词)	数据库
deploy	*vt. & vi.* (动词)	使展开

---

[39] 模拟
[40] 仪器仪表
[41] 原型
[42] 关键的

dictate	vt. & vi.（动词）	支配
graphically	adv.（副词）	生动地；用图表表示；用图解法
hierarchy	n.（名词）	层次
implementation	n.（名词）	实施；履行
incremental	n.（名词）	增量
instrumentation	n.（名词）	仪器仪表
intensive	adj.（形容词）	集中的；加强的
intervention	vt. & vi.（动词）	干预
iterative	adj.（形容词）	迭代
mapping	n.（名词）	映射
mechanism	n.（名词）	机制
migration	vt. & vi.（动词）	迁移
minimal	adj.（形容词）	最小的
multiple	n.（名词）	倍数
optimally	adv.（副词）	最佳
persistent	n.（名词）	持久性
precise	adj.（形容词）	精确
prototype	n.（名词）	原型
representation	n.（名词）	代表
semantics	n.（名词）	语义
simulation	n.（名词）	模拟
specify	n.（名词）	详述
transcend	vt. & vi.（动词）	超越
trivial	adj.（形容词）	琐碎；细小
unambigously	复合词	毫不含糊地
yield	vt. & vi.（动词）	产生

Phrases

a white board	一个白板
enterprise information systems	企业信息系统
error-prone	易出错
real time embedded systems	实时嵌入式系统
software-intensive system	软件-密集的系统
Unified Modeling Language	统一建模语言
Web-based applications	网页化应用程序

| well-defined | 定义明确的；界限清楚的 |

批 注

注 1　Learning to apply the UML effectively 为主语；starts 为谓语；which 引导一个非限制性定语从句；that dictate how these building blocks may be put together 的先行词为 the rules；that apply throughout the language 的先行词为 mechanisms。这句话的含义是：要学习使用 UML，一个有效的出发点是形成该语言的概念模型，这要求学习 3 个要素：UML 的基本构造块、支配这些构造块如何放置在一起的规则以及运用于整个语言的一些公共机制。

注 2　A well-defined process 为主语；will guide 为谓语；what，what and how 的先行词为 deciding。

注 3　这句话的含义是：他们甚至可以在白板或餐巾纸上草拟出一些想法。然而，这样做存在几个问题。第一，别人对这些概念模型容易产生错误的理解，因为并不是每个人都使用相同的语言。

注 4　这句话的含义是：例如，对于一个类的所有代码，虽可推断出它的含义，但不能直接领会它。类似地，在基于 Web 的系统中研究系统的代码虽可推断出对象的物理分布和可能的迁移，但也不能直接领会它。

注 5　more than：只不过，充其量。这句话的含义是：UML 只不过是图形符号的一个分支。

注 6　that 引导一个宾语从句，先行词是 means。or even to tables in a relational database or the persistent store of an object-oriented database. 这句前面省略了 it is possible。这句话的含义是：这意味着一种可能性，即可把用 UML 描述的模型映射成编程语言，如 Java、C++ 和 Visual Basic 等，甚至映射成关系数据库的表或面向对象数据库的永久储存。

注 7　这句话的含义是：可以进行逆向工程——由编程语言代码重新构造 UML 模型。

注 8　主语：Combining these two paths of forward code generate and reverse engineering yields round-trip engineering；谓语：mean；or a textual view 省略了 work in。这句话的含义是：把正向代码生成和逆向工程这两种方式结合起来就可以产生双向工程，这意味着既能在图形视图下工作，又能在文字视图下工作，只要用工具保持二者的一致性即可。

Associated Reading

Guidelines of UML Activity Diagrams

In many ways UML Activity diagrams are the object-oriented equivalent of flow charts and data-flow diagrams (DFDs). They are used to explore the logic of:

- A complex operation
- A complex business rule
- A single use case
- Several use cases
- A business process
- Software processes

Guidelines:

1. General guidelines

Place "The Start Point" In The Top-Left Corner. A start point is modeled with a filled in circle, using the same notation that UML State Chart diagrams use. Every UML Activity Diagram should have a starting point, and placing it in the top-left corner reflects the way that people in Western cultures begin reading. Figure 4-2 which models the business process of enrolling in a university, takes this approach.

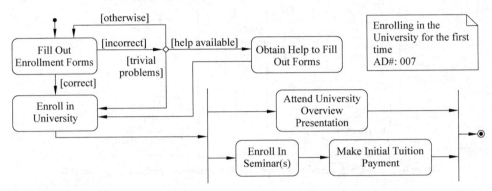

Figure 4-2   Activity diagram

Always Include an Ending Point. An ending point is modeled with a filled in circle with a border around it, using the same notation that UML State Chart diagrams use. Figure 4-2 is interesting because it does not include an end point because it describes a continuous process-sometimes the guidelines don't apply.

Flowcharting Operations Implies the Need to Simplify. A good rule of thumb is that if an operation is so complex you need to develop a UML Activity diagram to understand it that you should consider refectory it.

2. Decision points

A decision point is modeled as a diamond on a UML Activity diagram. Decision Points Should Reflect the Previous Activity. In Figure 4-2 you see that there is no label on the decision point, unlike traditional flowcharts which would include text describing the actual decision being made, you need to imply that the decision concerns whether the person was enrolled in the university based on the activity that the decision point follows. The guards depicted using the format [description], on the transitions leaving the decision point also help to describe the decision point.

Avoid Superfluous Decision Points. The Fill Out Enrollment Forms activity in Figure 4-2 includes an implied decision point, a check to see that the forms are filled out properly, which simplified the diagram by avoiding an additional diamond.

3. Swim lane guidelines

A swim lane is a way to group activities performed by the same actor on an activity diagram or to group activities in a single thread. Figure 4-3 includes three swim lanes, one for each actor.

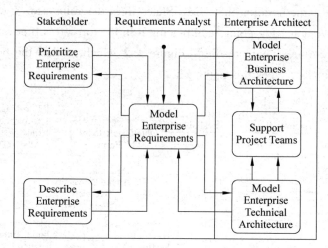

Figure 4-3　A UML activity diagram for the enterprise architectural modeling

Unit 5　Database

Text A　MySQL Introduction

The MySQL database system uses a **client/server❶ architecture❷** that centers on the server. The server is the program that actually **manipulates❸** databases. Client programs don't do that directly. Instead, they communicate your **intent❹** to the server by means of **statements❺** written in **Structured Query Language❻** (SQL). Client programs are installed locally on the machine from which you want to access MySQL, but the server can be installed anywhere, as long as clients can connect to it. 注1 MySQL is an **inherently❼** networked database system, so clients can communicate with a server that is running locally on your machine or one that is running somewhere else, perhaps on a machine **on the other side❽** of the planet. Clients can be written for many different purposes, but each **interacts❾** with the server by connecting to it, sending SQL statements to it to have database operations performed, and receiving the statement results from it.

One such client is the MySQL program that is included in MySQL **distributions❿**. When used interactively, MySQL prompts you for a statement, sends it to the MySQL server for execution, and then **displays⓫** the results. 注2 This capability makes MySQL useful in its own right, but it's also a valuable tool to help you with your MySQL programming activities. It's often **convenient to⓬** be able to quickly review the structure of a table that you're accessing from within a **script⓭**, to try a statement before using it in a program to make sure that it produces the right kind of output, and **so forth⓮**. MySQL is just right for these jobs. MySQL also can be used non-interactively; for example, to read statements from a file or

---

❶ 客户端/服务器
❷ 体系,模式
❸ 操作;操纵
❹ 意图;目的
❺ 语句,命令
❻ 结构化查询语言
❼ 与生俱来地
❽ 另一侧,另一端
❾ 互相影响
❿ 分配;分布
⓫ 展示;显示
⓬ 便捷的;方便的
⓭ 手迹;脚本
⓮ 等等

· 63 ·

from other programs. This enables you to use MySQL from within scripts or **cron**⑮ jobs or in **conjunction**⑯ with other applications.

This chapter describes MySQL's capabilities so that you can use it more effectively:

- Starting and stopping MySQL
- Specifying connection **parameters**⑰ and using option files
- Setting your PATH variable so that your command **interpreter**⑱ can find MySQL (and other MySQL programs)
- Issuing SQL statements **interactively**⑲ and using batch files
- Canceling and editing statements
- **Controlling**⑳ MySQL output format

To use the examples shown, you'll need a MySQL user account and a database to work with. The first two sections of the chapter describe how to use MySQL to set these up. For **demonstration**㉑ purposes, the examples **assume**㉒ that you'll use MySQL as follows:

- The MySQL server is running on the local host
- Your MySQL username and password are cbuser and cbpass
- Your database is named cookbook

For your own **experimentation**㉓, you can **violate**㉔ any of these assumptions. Your server need not be running locally, and you need not use the username, password, or database name that are used in this charpter.注3 Naturally, if you use different default values on your system, you'll need to change the examples accordingly.

Even if you do not use cookbook as the name of your database, I recommend that you create a database to be **dedicated**㉕ specifically to trying the examples shown here, rather than trying them with a database that you're using currently for other purposes.注4 Otherwise, the names of your existing tables may conflict with those used in the examples, and you'll have to make modifications to the examples that are unnecessary when you use a separate database.

If you have another favorite client program to use for issuing queries, some of the concepts covered in this chapter may not apply. For example, you might prefer the graphical MySQL Query Browser program, which provides a **point-and-click interface to**㉖

---

⑮ 时钟守护进程
⑯ 同时发生
⑰ 参数;系数
⑱ 解释器
⑲ 交互式地
⑳ 控制;管理
㉑ 示范;演示
㉒ 假定
㉓ 实验;试验
㉔ 违反
㉕ 专用的
㉖ 单击鼠标接口

MySQL databases. In this case, some of the principles will be different, such as the way that you **terminate**[27] SQL statements. In MySQL, you terminate statements with semicolon (;) characters, whereas in MySQL Query **Browser**[28] there is an Execute button for terminating statements.[注5] Another popular interface is phpMyAdmin, which enables you to access MySQL through your web browser.

Words

architecture	n.（名词）	风格,模式
assume	vt. & vi.（动词）	假定
browser	n.（名词）	浏览器
conjunction	n.（名词）	同时发生
controlling	vt. & vi.（动词）	控制;管理
dedicated	adj.（形容词）	专用的
demonstration	n.（名词）	示范;演示
display	vt. & vi.（动词）	展示;显示
distribution	n.（名词）	分配;分布
experimentation	n.（名词）	实验;试验
inherently	adv.（副词）	与生俱来地
intent	n.（名词）	意图;目的
interact	vt. & vi.（动词）	互相影响
interactively	adv.（副词）	交互式地
interpreter	n.（名词）	解释器
manipulate	vt. & vi.（动词）	操作;操纵
parameter	n.（名词）	参数;系数
script	n.（名词）	手迹;脚本
statement	n.（名词）	语句,命令
terminate	vt. & vi.（动词）	结束,终止
violate	vt.（动词）	违反

Phrases

client/server	客户端/服务器
convenient to	便捷的,方便的
on the other side	另一侧,另一端

---

[27] 结束,终止
[28] 浏览器

point-and-click interface to	单击鼠标接口
so forth	等等
Structured Query Language	结构化查询语言

Exercises

【Ex1】 Fill in the blanks according to the text:

(1) The MySQL database _____ uses a client-server architecture that centers around the server.

(2) The server is the program that actually _____ databases.

(3) This _____ makes MySQL useful in its own right, but it's also a valuable tool to help you with your MySQL programming activities.

(4) Otherwise, the names of your existing tables may _____ with those used in the examples.

(5) In this case, some of the principles will be different, such as the way that you _____ SQL statements.

【Ex2】 Translate into Chinese:

(1) The MySQL database system uses a client/server architecture that centers around the server.

(2) The server is the program that actually manipulates databases. Client programs don't do that directly.

(3) MySQL is an inherently networked database system, so clients can communicate with a server that is running locally on your machine or one that is running somewhere else.

(4) One such client is the MySQL program that is included in MySQL distributions.

(5) MySQL also can be used noninteractively; for example, to read statements from a file or from other programs.

(6) This enables you to use MySQL from within scripts or cron jobs or in conjunction with other applications.

(7) Another popular interface is phpMyAdmin, which enables you to access MySQL through your web browser.

(8) In this case, some of the principles will be different, such as the way that you terminate SQL statements.

【Ex3】 Choose the best answer:

(1) What's the system used in MySQL?

　　A. The MySQL database system uses a client/server architecture.

　　B. The MySQL database system uses a point-to-point architecture.

　　C. The MySQL database system uses both client/server and point-to-point architecture.

　　　　D. Not mentioned.

（2）Which part does actually manipulate the MySQL database?

　　　　A. The server.　　　　　　　　　B. The client.

　　　　C. The sever and client.　　　　　D. Not mentioned.

（3）Where could a client program was installed?

　　　　A. Locally on your machine.　　　B. Other place where have net.

　　　　C. The mechine near the server.　D. Not mentioned.

（4）Which is begond MySQL's capabilities?

　　　　A. Starting and stopping MySQL.

　　　　B. Canceling and editing statements.

　　　　C. Specifying connection parameters and using option files.

　　　　D. Manipulating the MySQL database.

（5）Where does the MySQL execute statement?

　　　　A. In MySQL.　　　　　　　　　B. In client.

　　　　C. In server.　　　　　　　　　　D. Both in client and server.

批　注

　　注1　主语：Client programs；谓语：are；but the server can be installed anywhere, as long as clients can connect to it 与前面是并列关系；as long as 引导一个条件状语从句。

　　注2　这句话的含义是：当你交互时，MySQL 通过一个语句提示把语句发送到 MySQL 服务器执行，然后显示结果。

　　注3　这句话是一个并列句；Your server, and you 为句子的主语；that are used in this chapter 引导一个定语从句，先行词是 username, password, or database name。这句话的含义是：你的服务器可能不在本地运行，你不需要使用用户名、密码、数据库的名字。

　　注4　这句话的含义是：我建议你根据这里提供的例子创建一个数据库，或者为不同的目的试用它们。

　　注5　whereas 引导一个非限制性定语从句。这句话的含义是：在 MySQL 里，你可以使用分号作为结束语句，而 MySQL Query Browser 中有一个专门的执行符号用来结束语句。

Text B　Performing Transactions

　　The MySQL server can **handle**❶ **multiple**❷ clients at the same time because it is **multithreaded**❸. To **deal with**❹ contention among clients, the server performs any

---

❶ 处理；操作
❷ 并联，多个的
❸ 多线程的
❹ 处理

necessary locking **so that**❺ two clients cannot modify the same data at once.注1 However, as the server executes SQL statements, it's very possible that **successive**❻ statements received from a given client will be **interleaved**❼ with statements from other clients. If a client **issues**❽ multiple statements that are **dependent on**❾ each other, the fact that other clients may be updating tables in between those statements can cause difficulties. Statement failures can be **problematic**❿, too, if a multiple-statement operation does not run to completion. Suppose that you have a flight table containing information about airline flight schedules and you want to update the row for Flight 578 by choosing a pilot from among those available.注2 You might do so using three statements as follows:

```
SELECT @ p_val : = pilot_id FROM pilot WHERE available = 'yes' LIMIT 1;
UPDATE pilot SET available = 'no' WHERE pilot_id = @ p_val;
UPDATE flight SET pilot_id = @ p_val WHERE flight_id = 578;
```

The first statement chooses one of the available pilots, the second marks the pilot as unavailable, and the third **assigns**⓫ the pilot to the flight.注3 That's straightforward enough in practice, but **in principle**⓬ there are **a couple of**⓭ significant difficulties with the process:

If two clients want to schedule pilots, it's possible that both of them would run the **initial**⓮ SELECT query and retrieve the same pilot ID number before either of them has a chance to set the pilot's status to unavailable. If that happens, the same pilot would be scheduled for two flights at once.

All three statements must execute successfully as a unit. For example, if the SELECT and the first UPDATE run successfully, but the second UPDATE fails, the pilot's status is set to unavailable without the pilot being assigned a flight. The database will be left in an inconsistent state.

To prevent concurrency and integrity problems in these types of situations, **transactions**⓯ are helpful. A transaction groups **a set of**⓰ statements and **guarantees**⓱ the following **properties**⓲:

---

❺ 以使,以便
❻ 连续的
❼ [计]交错
❽ 发送
❾ 依赖
❿ 有疑问的
⓫ 分配;指派
⓬ 原则上
⓭ 许多
⓮ 最初的;字首的
⓯ 处理,事务
⓰ 一组
⓱ 保证;担保
⓲ 性能;属性

- No other client can update the data used in the transaction while the transaction is in progress; For example, other clients cannot modify the pilot or flight records while you're booking a pilot for a flight. By preventing other clients from interfering with the operations you're performing, transactions solve concurrency problems arising from the multiple-client nature of the MySQL server. Transactions **serialize**[19] access to a shared resource across multiple-statement operations.
- Statements in a transaction are grouped and are **committed**[20] (take effect) as a unit, but only if they all succeed.[注4] If an error **occurs**[21], any actions that occurred **prior**[22] to the error are rolled back, leaving the **relevant**[23] tables unaffected as though none of the statements had been issued at all. This keeps the database from becoming **inconsistent**[24]. For example, if an update to the flights table fails, rollback causes the change to the pilots table to be undone, leaving the pilot still available. Rollback frees you from having to **figure out**[25] how to undo a partially completed operation yourself.

Words

assign	*vt.* & *vi.* (动词)	分配;指派
committed	*adj.* (形容词)	提交;起作用
guarantee	*vt.* & *vi.* (动词)	保证;担保
handle	*vt.* & *vi.* (动词)	处理;操作
inconsistent	*adj.* (形容词)	不一致的
initial	*adj.* (形容词)	最初的;字首的
interleave	*adj.* (形容词); *v.* (动词)	隔行扫描的交错;交叉存取
issue	*vt.* & *vi.* (动词)	发布
multiple	*adj.* (形容词)	并联,多个的
multithread	*adj.* (形容词)	多线程的
occur	*vt.* & *vi.* (动词)	发生;出现
prior	*adj.* (形容词)	之前的
problematic	*adj.* (形容词)	有疑问的
property	*n.* (名词)	性能;属性

---

[19] 串行序列地
[20] 提交;起作用
[21] 发生;出现
[22] 之前的
[23] 有关的
[24] 不一致的
[25] 找出,计算出

relevant	adj.（形容词）	有关的
serialize	vt. & vi.（动词）	串行序列地
successive	adj.（形容词）	连续的
transaction	n.（名词）	处理,事务

Phrases

a couple of	许多
a set of	一组
deal with	处理
dependent on	依赖
figure out	找出,计算出
in principle	大体上,原则上
so that	以便;以使

批 注

注1 这句话的含义是:为了处理客户端间内容方面的事务,服务器有一个锁的操作,这样两个客户端程序不可能在同一时刻修改相同数据。

注2 这句话的含义是:假设你有一张航班信息表,里面包含航线航班计划信息,需要从可供选择的飞行员中选择一位飞行员,更新578航班记录。

注3 这句话的含义是:第一条语句用于选择一个飞行员,第二条语句用于设置该飞行员为无效,第三条语句将该飞行员分配给这次航班。

注4 这句话的含义是:在事务里的语句是成组的,只有当它们执行成功时,是作为整个单元奏效的。

Associated Reading

Pattern Matching in SQL Server Queries

You may often need to create a SQL Server query that performs inexact pattern matching through the use of wildcard characters. The use of wildcards allows you to find data that fits a certain pattern, rather than specifying it exactly. For example, you can use the wildcard 'C%' to match any string beginning with a capital C.

To use a wildcard expression in a SQL query, you'll need to use the LIKE clause to specify it. For example, you could search for any employee in your database with a last name beginning with the letter C using the following SQL statement:

SELECT * FROM employees WHERE last_name LIKE 'C%'

There are several different wildcard expressions supported by Transact SQL:

- The % wildcard matches zero or more characters of any type. If you're **familiar with**[20] DOS pattern matching, it's the equivalent of the * wildcard in that syntax.
- The _ wildcard matches exactly one character of any type. It's the **equivalent of**[27] the ? wildcard in DOS pattern matching.
- You can specify a list of characters by enclosing them in square brackets. For example, the wildcard [aeiou] will match any vowel.
- You can specify **a range of**[28] characters by enclosing the range in square brackets. For example, the wildcard [a-m] will match any letter in the first half of the alphabet.
- You can negate a range of characters by including the carat character immediately inside of the opening square bracket. For example, [^aeiou] matches any non-vowel character while [^a-m] matches any character not in the first half of the alphabet.

You may also combine these wildcards **in complex patterns to**[29] perform more advanced queries. For example, suppose you needed to construct a list of all of your employees who have names that begin with a letter from the first half of the alphabet but do NOT end with a vowel. You could use the following query:

```
SELECT * FROM employees WHERE last_name LIKE '[a-m]%[^aeiou]'
```

Similarly, you could construct a list of all employees with last names consisting of exactly four characters by using the query:

```
SELECT * FROM employees WHERE last_name LIKE '____'
```

As you can tell, the use of SQL Server's pattern matching capabilities offers database users the ability to go beyond simple text queries and perform advanced searching operations.

---

[20] 熟悉
[27] 等同于
[28] 范围
[29] 复杂的形式

Unit 6　Embedded System

Text A　What is an Embedded System?

One of the more surprising developments of the last few decades has been the **ascendance**❶ of computers to a position of **prevalence**❷ in human affairs❸. Today there are more computers in our homes and offices than there are people who live and work in them. 注1 Yet many of these computers are not recognized as such by their users. In this unit, I'll explain what embedded systems are and where they are found. I will also introduce the subject of embedded programming, explain why I have selected C and C++ as the languages for this book, and describe the hardware used in the examples.

1. About embedded system

An *embedded system* is a combination of computer hardware and software, and perhaps additional mechanical or other parts, designed to perform a specific function. A good example is the **microwave oven**❹. Almost every household has one, and tens of millions of them are used every day, but very few people realize that a processor and software are **involved in**❺ the **preparation**❻ of their lunch or dinner.

This is in direct **contrast**❼ to the personal computer in the family room. It too is **comprised of**❽ computer hardware and software and mechanical **components**❾ (disk drives, for example). However, a personal computer is not designed to perform a specific function. Rather, it is able to do many different things. Many people use the term ***general-purpose computer***❿ to make this **distinction**⓫ clear. As **shipped**⓬, a general-purpose

---

❶ 优势；支配力量
❷ 传播，流行，普及
❸ 人类事务
❹ 微波炉
❺ 参与
❻ 准备，预备，配置
❼ 对比，反差
❽ 由……组成
❾ 部件，组成部分
❿ 通用计算机
⓫ 差别，分别；优秀，荣誉
⓬ 装船

computer is a blank **slate**⑬; the **manufacturer**⑭ does not know what the customer will do with it. One customer may use it for a network file server, another may use it **exclusively**⑮ for playing games, and a third may use it to write the next great American novel.

Frequently, an embedded system is a component within some larger system. For example, modern cars and trucks contain many embedded systems. One embedded system controls the **anti-lock brakes**⑯, another **monitors**⑰ and controls the **vehicle's**⑱ **emissions**⑲, and a third displays information on the **dashboard**⑳.注2 In some cases, these embedded systems are connected by some sort of a communications network, but that is certainly not a requirement.

At the possible risk of confusing you, it is important to point out that a general-purpose computer is itself **made up of**㉑ **numerous**㉒ embedded systems. For example, my computer consists of a keyboard, mouse, video card, modem, hard drive, floppy drive, and sound card-each of which is an embedded system. Each of these devices contains a processor and software and is designed to perform a specific function. For example, the modem is designed to send and receive digital data over an **analog**㉓ telephone line. That's it. And all of the other devices can be **summarized**㉔ in a single sentence as well.

If an embedded system is designed well, the existence of the processor and software could be completely **unnoticed**㉕ by a user of the device. Such is the case for a microwave oven, VCR, or alarm clock. In some cases, it would even be possible to build an **equivalent**㉖ device that does not contain the processor and software. This could be done by replacing the combination with a custom **integrated circuit**㉗ that performs the same functions in hardware. However, a lot of flexibility is lost when a design is hard-coded in this way. It is much easier, and cheaper, to change a few lines of software than to redesign **a piece of**㉘ custom hardware.

⑬ 板岩,石板
⑭ 制造商,制造厂
⑮ 专门的,排除其他的
⑯ 防抱死制动系统
⑰ 监视器,显示器
⑱ 交通工具,车辆
⑲ 发出,发光,放射物
⑳ 仪表盘
㉑ 由……组成
㉒ 众多的,许多的
㉓ 模拟物,模拟
㉔ 概括,总结
㉕ 被忽略的,不被重视的
㉖ 等价物
㉗ 集成电路
㉘ 一片,一块

2. History and future

The first such systems could not possibly have appeared before 1971. That was the year Intel introduced the world's first **microprocessor**㉙. This **chip**㉚, the 4004, was designed for use in a line of business calculators produced by the Japanese company Busicom. In 1969, Busicom asked Intel to design a set of custom integrated circuits-one for each of their new calculator models. The 4004 was Intel's response. Rather than design custom hardware for each calculator, Intel proposed a general-purpose circuit that could be used throughout the entire line of calculators. This general-purpose processor was designed to read and execute a set of instructions-software-stored in an **external**㉛ memory chip. Intel's idea was that the software would give each calculator its unique set of features.

The microprocessor was an overnight success, and its use increased steadily over the next decade. Early embedded applications included **unmanned**㉜ space **probes**㉝, **computerized**㉞ traffic lights, and aircraft flight control systems. In the 1980s, embedded systems quietly rode the waves of the microcomputer age and brought microprocessors into every part of our personal and professional lives. Many of the electronic devices in our kitchens (**bread machines**㉟, **food processors**㊱, and microwave ovens), living rooms (televisions, stereos, and remote controls), and workplaces (fax machines, pagers, laser printers, cash registers, and credit card readers) are embedded systems.

It seems **inevitable**㊲ that the number of embedded systems will continue to increase rapidly. Already there are promising new embedded devices that have enormous market **potential**㊳: light switches and **thermostats**㊴ that can be controlled by a central computer, intelligent **air-bag**㊵ systems that don't inflate when children or small adults are present, palm-sized **electronic organizers**㊶ and **personal digital assistants** (**PDAs**㊷), digital cameras, and dashboard **navigation**㊸ systems.注3 Clearly, individuals who possess the skills

---

㉙ 微处理器
㉚ 芯片
㉛ 外部的
㉜ 无人操纵的,未经过训练的
㉝ 探索,调查
㉞ 用计算机操作的
㉟ 面包机
㊱ 食品处理器
㊲ 不可避免的,必然发生的
㊳ 潜力
㊴ 恒温器
㊵ 气囊
㊶ 电子组织者
㊷ 个人数字助理
㊸ 导航,航行,航海

and desire to design the next generation of embedded systems will be in demand for quite some time.

3. Real-Time systems

One subclass of embedded systems **is worthy of**⑭ an **introduction**⑮ at this point. As commonly defined, a real-time system is a computer system that has timing **constraints**⑯. In other words, a real-time system is partly specified **in terms of**⑰ its ability to make certain calculations or decisions in a timely manner. These important calculations are said to have deadlines for completion. And, for all practical purposes, a missed deadline is just **as bad as**⑱ a wrong answer.

The issue of what happens if a deadline is missed is a crucial one. For example, if the real-time system is part of an airplane's flight control system, it is possible for the lives of the passengers and crew to be endangered by a single missed deadline. However, if instead the system **is involved in**⑲ **satellite**⑳ communication, the damage could be limited to a single **corrupt**㉑ data packet. 注4 The more severe the consequences, the more likely it will be said that the deadline is "hard" and, thus, the system a hard real-time system. Real-time systems **at the other end of**㉒ this **continuum**㉓ are said to have "soft" deadlines.

The designer of a real-time system must be more diligent in his work. He must guarantee reliable operation of the software and hardware under all possible conditions. 注5 And, to the degree that human lives depend upon the system's proper execution, this guarantee must be backed by engineering calculations and **descriptive**㉔ paperwork.

Words

analog	n.（名词）; adj.（形容词）	[自]模拟;[自]模拟的;有长短针的,类比
anti-lock brakes	复合名词	防刹车锁死
ascendance	n.（名词）	优势;支配力量

---

⑭ 值得
⑮ 介绍,引进
⑯ 约束,限制
⑰ 根据,就……而言
⑱ 与……一样糟糕
⑲ 参与
⑳ 微型,人造卫星
㉑ 破坏的,腐败的
㉒ 另一方面
㉓ 连续时间
㉔ 描述的,叙述的

chip	*n.*（名词）	芯片
component	*n.*（名词）	部件,组成部分
computerize	*adj.*（形容词）	用计算机操作的
constraint	*n.*（名词）	约束,限制
contrast	*vt. & vi.*（动词）	对比,反差
corrupt	*adj.*（形容词）	破坏的,腐败的
dashboard	*n.*（名词）	仪表盘
descriptive	*adj.*（形容词）	描述的,叙述的
distinction	*n.*（名词）	差别,分别;优秀,荣誉
emission	*n.*（名词）	发出,发光,放射物
equivalent	*n.*（名词）	等价物
exclusively	*adv.*（动词）	仅,占
external	*adj.*（形容词）	外部的
inevitable	*adj.*（形容词）	不可避免的,必然发生的
introduction	*n.*（名词）	介绍,引进
manufacturer	*n.*（名词）	制造商,制造厂
microprocessor	*n.*（名词）	微处理器
microwave	*n.*（名词）	微波,微波炉
monitor	*n.*（名词）	检测器,显示器
navigation	*n.*（名词）	导航,航行,航海
numerous	*adj.*（形容词）	众多的,许多的
PDA	*n.*（名词）	个人数字助理
potential	*n.*（名词）	电位,电势
preparation	*n.*（名词）	准备,预备,配置
prevalence	*n.*（名词）	传播,流行,普及
probe	*vt. & vi.*（动词）	探索,调查
satellite	*n.*（名词）	微型,人造卫星
shipped	*vt. & vi.*（动词）	装船（ship 的过去分词）;发货
slate	*n.*（名词）	板岩,石板
summarize	*vt. & vi.*（动词）	概括,总结
thermostat	*n.*（名词）	恒温器
unmanned	*adj.*（形容词）	无人操纵的,未经过训练的
unnoticed	*adj.*（形容词）	被忽略的,不被重视的
vehicle	*n.*（名词）	交通工具,车辆

Phrases

a piece of	一片,一块
air-bag	气囊

as bad as	与……一样糟糕
at the other end of	另一方面
be involved in	参与
be worthy of	值得,配得上
bread machines	面包机
comprised of	由……组成
continuum	连续时间
electronic organizers	电子组织者
food processors	食品处理器
general-purpose computer	通用计算机
in human affairs	人类事务
in terms of	依据;就……而言
integrated circuit	集成电路
involved in	涉及;包含;牵涉进……
made up of	由……组成
microwave ovens	微波;微波炉
personal digital assistants	个人数字助理

Exercises

【Ex1】 Fill in the blanks according to the text:

Early embedded applications included _____ space _____, _____ traffic lights, and aircraft _____ control systems. In the 1980s, embedded systems quietly rode the waves of the microcomputer age and brought microprocessors into every part of our personal and professional _____.

【Ex2】 Translate into Chinese:

(1) An *embedded system* is a combination of computer hardware and software, and perhaps additional mechanical or other parts, designed to perform a specific function.

(2) In the 1980s, embedded systems quietly rode the waves of the microcomputer age and brought microprocessors into every part of our personal and professional lives.

(3) For example, if the real-time system is part of an airplane's flight control system, it is possible for the lives of the passengers and crew to be endangered by a single missed deadline.

(4) The hardware reads digital data from one set of electrical connections and writes an analog version of the data to an attached telephone line.

(5) If you are lucky, the documentation provided with your hardware will contain a

superset of the block diagram you need.

(6) If you think about it from this perspective, one thing you quickly realize is that the processor has a lot of compatriots.

(7) Instead of simply storing the data that is provided to it, a peripheral might instead interpret it as a command or as data to be processed in some way.

(8) Already there are promising new embedded devices that have enormous market potential: light switches and thermostats that can be controlled by a central computer, intelligent air-bag systems that don't inflate when children or small adults are present, palm-sized electronic organizers and personal digital assistants (PDAs), digital cameras, and dashboard navigation systems.

【Ex3】 Choose the best answer:

(1) Today there are _____ computers in our homes and offices than there are people who live and work in them.

  A. less   B. more   C. little   D. lots of

(2) For example, the modem is designed to send and receive digital data _____ an analog telephone line.

  A. over   B. in   C. through   D. on

(3) Rather than design custom hardware for each calculator, Intel proposed a general-purpose _____ that could be used throughout the entire line of calculators.

  A. computer   B. board   C. machine   D. circuit

(4) Before picking _____ the board, you should be able to answer two basic questions about it.

  A. out   B. up   C. on   D. to

(5) Data to be printed is _____ from either serial port, held in RAM until the printer is ready for more data, and delivered to the printer via the parallel port.

  A. trended   B. transported   C. accepted   D. received

批 注

注1 more… than：比……更。这句话的含义是：今天在我们的家里和办公室里，计算机的数量要比使用它们生活和工作的人还多。

注2 这句话的含义是：一个嵌入式系统会被用来控制防抱死制动系统，另一个用来监控车辆的气体排放，还有的用于在仪表板上显示信息。

注3 这句话的含义是：已经有很多具有巨大市场潜力的新的嵌入式设备了：可以被中央计算机控制的调光器和恒温器、当小孩子或矮个子的人在的时候不会充气的智能气囊、掌上电子记事簿和个人数字助理（PDA）、数码照相机和仪表导航系统。

注4 这句话的含义是：如果一个实时系统是飞机飞行控制系统的一部分，那么一个超期的计算就可能使乘客和机组人员的生命受到威胁。而把这个系统用在卫星通信环境下，危害也许可以限制在仅

一个损坏的数据包。

注5　这句话的含义是：他必须保证软件和硬件在所有可能的情况下都能可靠工作。

Text B　Getting to Know the Hardware

As an embedded software engineer, you'll have the opportunity to work with many different pieces of hardware in your career. In this chapter, I will teach you a simple procedure that I use to **familiarize** myself **with**❶ any new board. In the process, I'll guide you through the creation of a header file that describes the board's most important features and a piece of software that initializes the hardware to a known state.^{注1}

1. Understand the big picture

Before writing software for an embedded system, you must first be familiar with the hardware on which it will run. At first, you just need to understand the general operation of the system. You do not need to understand every little detail of the hardware; that kind of knowledge will not be needed right away and will come with time.

Whenever you receive a new board, you should take some time to read whatever **document**❷ have been provided with it. If the board is an off-the-shelf product, it might arrive with a "User's Guide" or "Programmer's Manual" that has been written with the software developer in mind.^{注2} However, if the board was custom designed for your project, the documentation might be more **cryptic**❸ or written mainly for the **reference**❹ of the hardware designers. Either way, this is the single best place for you to start.

While you are reading the documentation, set the board itself aside. This will help you to focus on the big picture. There will **be plenty of**❺ time to examine the actual board more closely when you have finished reading. Before picking up the board, you should be able to answer two basic questions about it:

- What is the overall purpose of the board?
- How does data flow through it?

For example, imagine that you are a member of a modem design team. You are a software developer who has just received an early **prototype**❻ board from the hardware designers. Because you are already familiar with modems, the overall purpose of the board and the data-flow through it should be fairly obvious to you. The purpose of the board is to

---

❶　熟悉
❷　公文,文件,证件
❸　神秘的
❹　提到,参阅
❺　丰富,充足,大量
❻　原型

send and receive digital data over an analog telephone line. The hardware reads digital data from one set of electrical connections and writes an **analog**❼ version of the data to an attached telephone line.注3 Data also flows in the opposite direction, when analog data is read from the telephone line jack and output digitally.

Though the purpose of most systems is fairly obvious, the flow of the data might not be. I often find that a data-flow diagram is helpful in achieving rapid **comprehension**❽. If you are lucky, the documentation provided with your hardware will contain a superset of the block diagram you need. However, you might still find it useful to create your own data-flow diagram. That way, you can leave out those hardware components that are unrelated to the basic flow of data through the system.

In the case of the **Arcom board**❾, the hardware was not designed with a particular application in mind. So for the remainder of this chapter, we'll have to imagine that it does have a purpose. We shall assume the board was designed for use as a **printer-sharing**❿ device. A printer-sharing device allows two computers to share a single printer.注4

The user of the device connects one computer to each serial port and a printer to the **parallel**⓫ port.注5 Both computers can then send documents to the printer, though only one of them can do so at a given time.

In order to **illustrate**⓬ the flow of data through the printer-sharing device, I've drawn the diagram in Figure 6-1. (Only those hardware devices that are involved in this application of the Arcom board are shown.) By looking at the block diagram, you should be able to quickly visualize the flow of the data through the system. Data to be printed is accepted from either serial port, held in RAM until the printer is ready for more data, and delivered to the printer via the parallel port. The software that makes all of this happen is stored in ROM.

Once you've created a block diagram, don't just **crumple** it **up**⓭ and **throw** it **away**⓮. You should instead put it where you can refer to it throughout the project. I recommend creating a project notebook or binder with this data-flow diagram on the first page. As you continue working with this piece of hardware, write down everything you learn about it in your notebook. You might also want to keep notes about the software design and implementation. A project notebook is valuable not only while you are developing the software, but also once the project is complete.注6 You will **appreciate**⓯ the extra effort you

---

❼ 类似物,模拟
❽ 理解力
❾ Arcom 公司生产的板
❿ 打印机共享
⓫ 平行的,并行的
⓬ 说明,阐明
⓭ 弄皱
⓮ 使扭曲,使崩溃
⓯ 赏识,为表示感谢

put into keeping a notebook when you need to make changes to your software, or work with similar hardware, months or years later.

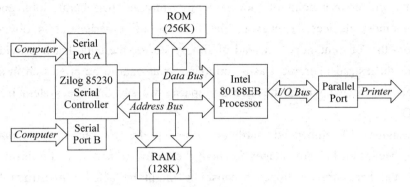

Figure 6-1　flow of data through the printer-sharing device

2. Examine the landscape

It is often useful to put yourself in the processor's shoes for a while. <u>After all, the processor is only going to do what you **ultimately**⑯ instruct it to do with your software.</u> 注7 Imagine what it is like to be the processor: what does the processor's world look like? If you think about it from this **perspective**⑰, one thing you quickly realize is that the processor has a lot of **compatriots**⑱. These are the other pieces of hardware on the board, with which the processor can communicate directly. In this section you will learn to recognize their names and addresses.

The first thing to notice is that there are two basic types: memories and peripherals. <u>Obviously, memories are for data and code storage and **retrieval**⑲.</u> 注8 But you might be wondering what the peripherals are. These are specialized hardware devices that either **coordinate interaction with**⑳ the outside world (I/O) or perform a specific hardware function. <u>For example, two of the most common peripherals in embedded systems are serial ports and timers. The former is an I/O device, and the latter is basically just a counter.</u> 注9

Members of Intel's 80x86 and some other processor families have two distinct address spaces through which they can communicate with these memories and peripherals. The first address space is called the memory space and is intended mainly for memory devices; the second is reserved **exclusively**㉑ for peripherals and is called the I/O space. However,

---

⑯　最后,最终
⑰　视角,观点,远景
⑱　同胞、伙伴
⑲　检索
⑳　协调相互作用
㉑　专门地,排除其他的

peripherals can also be located within the memory space, at the discretion of the hardware designer. When that happens, we say that those peripherals are **memory-mapped**[22].

From the processor's point of view, memory-mapped peripherals look and act very much like memory devices. However, the function of a peripheral is obviously quite different from that of a memory. Instead of simply storing the data that is provided to it, a peripheral might instead interpret it as a command or as data to be processed in some way. If peripherals are located within the memory space, we say that the system has memory-mapped I/O.

The designers of embedded hardware often prefer to use memory-mapped I/O exclusively, because it has advantages for both the hardware and software developers. It is attractive to the hardware developer because he might be able to **eliminate**[23] the I/O space, and some of its associated wires, altogether. This might not significantly reduce the production cost of the board, but it might reduce the **complexity**[24] of the hardware design. Memory-mapped peripherals are also better for the programmer, who is able to use pointers, data structures, and unions to interact with the peripherals more easily and efficiently.

Words

appreciate	vt. & vi.（动词）	赏识,为表示感谢
compatriot	n.（名词）	同胞;伙伴
complexity	n.（名词）	复杂,复杂性
comprehension	n.（名词）	理解力
coordinate	n.（名词）	坐标,同等的人
crumple	vt. & vi.（动词）	使扭曲,使崩溃
cryptic	adj.（形容词）	神秘的
document	n.（名词）	公文,文件,证件
eliminate	vt. & vi.（动词）	消除,根除,排除
exclusively	adv.（副词）	专门地,排除其他的
familiarize	vt. & vi.（动词）	熟悉
illustrate	v.（动词）	说明,阐明
parallel	adj.（形容词）	平行的,并行的
perspective	n.（名词）	视角,观点,远景
plenty	adj.（形容词）	丰富,充足,大量

---

[22] 内存映射
[23] 消除,根除,排除
[24] 复杂,复杂性

reference	n.（名词）	涉及,提及
retrieval	vt. & vi.（动词）	检索
ultimately	adv.（副词）	最后,最终
version	n.（名词）	版本;译文

Phrases

an imperative component of	不可或缺的组成部分
be familiar with	熟悉
be plenty of	很多
crumple up	弄皱
defined as	定义为
familiarize with	熟悉
interaction with	交往;与……相互作用
memory-mapped	内存映射的;存储映像的
printer-sharing	打印机共享
the General Public License	通用公共许可证的
throw away	扔掉
with the help of	在……的帮助下

批 注

注1 这句话的含义是:我将通过教你创建一个描述电路板的头文件和一小部分软件,在这个头文件里描述的是这张板的最重要特性,这一小部分软件把硬件初始化至一个已知状态。

注2 这句话的含义是:如果这个电路板是从货架上拿来的标准产品,那么它很可能会有"用户手册"或"程序员手册",里面刻记着软件开发商的名字。

注3 这句话的含义是:硬件从一组电源连接器上读取数字信号,然后转换成模拟信号,通过相连的电话线进行传输。

注4 这句话的含义是:一个共享的打印机设备可以允许两台计算机共享一个打印机。

注5 and a printer to the parallel port 省略了 connects,这句话是一个并列句,通过 and 将两句话连接起来。这句话的含义是:设备将一台计算机与每个串口相连,将打印机与并口相连。

注6 not only but also:不仅,而且;not only 后面连接了一个从句;but also 后面也连接了一个从句。这句话的含义是:项目工程记事本对于你开发一个软件很重要,对于项目开发完成后也是很重要的。

注7 这句话的含义是:毕竟,这个处理器只有当你通过软件发出指令做什么之后,才决定做什么。

注8 这句话的含义是:很显然,内存是用来存储数据、代码和检索的。

注9 the former 是指 serial ports;the latter 指的是 timers。这句话的含义是:例如,在嵌入式系统里有两个最常见的外设:串口和计时器。串口是输入输出设备,而计时器仅指计时器。

Associated Reading

Briet Introduction of Qt

Qt is defined as a cross platform application. It is a framework which is used in to develop application software. This application software is developed with the help graphical user interface. This cross platform application is also used in developing non graphical user interface programs. Some of the programs which use this application program are Google, Adobe Photoshop, Panasonic; Philips etc. This application uses the C++ language. But Qt uses a unique code generator with many macros to enhance the language. Qt can also be used in many programming language with the help of language bindings. It is used on many platforms and is also internationally recognized. This cross platform application was developed by Trolltech but they later sold this application to Nokia. The software libraries which use this application are advanced component framework, KDELibs, LibQxt etc.

Qt has always been available through a commercial license which allows the expansion of proprietary purposes without any restriction on licensing. In addition, Qt has been steadily made open through an increasing number of other free licenses. Currently, Qt is accessible under the General Public License (GNU); this provision makes it available for utilization for both free software and proprietary. Before the 1.45 version, the source code used for Qt was published under free Qt license. This was not really viewed as amenable with the free software description of the Free Software Foundation and neither by the Open Source principle as described by the Open Source Initiative. The reason behind this was that it did not allow the distribution of other modified versions.

In 1998, many controversies broke out when it was acknowledged that KDE's KDE software compilation would undoubtedly become the leading desktop setting especially for Linux. Owing to the fact that it was founded on a Qt basis, many professionals who took part in the free software movement became apprehensive that an imperative component of one of a leading operating system would become proprietary.

Along with the release of the 2.0 version of the toolbox, the license was transformed to the QPL (Q Public License). QPL was a free software license but it was considered incompatible with GPL by the Free Software Foundation. Trolltech and KDE then sought out many compromises which would imply that Qt will not be regarded by any license more restrictive than QPL, even in the case Trolltech goes bankrupt. This major issue led to the invention of the free Qt foundation by KDE.

Qt is defined as a cross platform application. It is a framework which is used in to develop application software. This application software is developed with the help graphical user interface. This cross platform application is also used in developing non graphical user interface programs. Some of the programs which use this application program are Google, Adobe Photoshop, Panasonic, Philips etc. This application uses the C++ language. But Qt uses a unique code generator with many macros to enhance the language. Qt can also be used in many programming language with the help of language bindings. It is used on many platforms and is also internationally recognised. This cross platform application was developed by Trolltech but they later sold this application to Nokia. The software libraries which use this application are advanced component framework, KDELibs, LibQxt etc.

Unit 7　Computer Network

Text A　Internet Protocol Suite

The Internet Protocol Suite is the set of communications **protocols**❶ used for the Internet and other similar networks. It is commonly also known as TCP/IP, named from two of the most important protocols in it: the **Transmission**❷ Control Protocol (TCP) and the Internet Protocol (IP), which were the first two networking protocols defined in this **standard**❸. Modern IP networking represents a **synthesis**❹ of several developments that began to **evolve**❺ in the 1960s and 1970s, namely the Internet and local area networks, which emerged during the 1980s, together with the **advent**❻ of the World Wide Web in the early 1990s. 注1

The Internet Protocol Suite, like many protocol suites, is **constructed**❼ as **a set of** ❽ layers. Each layer solves a set of problems involving the transmission of data. In particular, the layers define the **operational**❾ scope of the protocols within.

Often a component of a layer provides a **well-defined**❿ service to the upper layer protocols and may be using services from the lower layers. Upper layers are **logically**⓫ closer to the user and deal with more **abstract**⓬ data, relying on lower layer protocols to **translate**⓭ data into forms that can eventually be **physically**⓮ transmitted.

The TCP/IP model consists of four layers. From lowest to highest, these are the Link Layer, the Internet Layer, the Transport Layer, and the Application Layer.

1. History

The Internet Protocol Suite resulted from research and development conducted by the

---

❶ 协议
❷ 传输
❸ 标准
❹ 综合
❺ 演变;进化
❻ 出现;到来
❼ 组成
❽ 一组
❾ 可操作的,运行的
❿ 明确定义的
⓫ 逻辑上
⓬ 抽象的
⓭ 翻译,转换
⓮ 物理学的

Defense Advanced Research Projects Agency (DARPA) in the early 1970s. In the spring of 1973, Vinton Cerf, the developer of the **existing**⑮ ARPANET Network Control Program (NCP) protocol, joined Kahn to work on **open-architecture**⑯ **interconnection**⑰ models with the goal of designing the next protocol **generation**⑱ for the ARPANET.

By the summer of 1973, Kahn and Cerf had worked out a **fundamental**⑲ **reformulation**⑳, where the differences between network protocols were **hidden**㉑ by using a common internetwork protocol, and, **instead of**㉒ the network being responsible for **reliability**㉓, as in the ARPANET, the hosts became responsible.

In 1975, a two-network TCP/IP communications test was performed between Stanford and University College London (UCL). In November, 1977, a three-network TCP/IP test was conducted between sites in the US, UK, and Norway. Several other TCP/IP **prototypes**㉔ were developed at **multiple**㉕ research centers between 1978 and 1983. The **migration**㉖ of the ARPANET to TCP/IP was officially completed on January 1, 1983, when the new protocols were **permanently**㉗ **activated**㉘.

In March 1982, the US Department of Defense declared TCP/IP as the standard for all military computer networking. In 1985, the Internet Architecture Board held a three day workshop on TCP/IP for the computer industry, attended by 250 **vendor**㉙ representatives, **promoting**㉚ the protocol and leading to its increasing commercial use.注2

2. Layers in the Internet protocol suite

The TCP/IP suite uses **encapsulation**㉛ to provide abstraction of protocols and services. Such encapsulation usually is **aligned**㉜ with the division of the protocol suite into layers of general functionality.注3 In general, an application (the highest level of the model)

---

⑮ 现有的
⑯ 开放式体系结构
⑰ 互相联络
⑱ 产生;一代
⑲ 基本的
⑳ 再形成;重塑
㉑ 难以发现的,隐藏
㉒ 取代;不是
㉓ 可靠性
㉔ 原型
㉕ 多重的
㉖ 迁移
㉗ 永存地,不变地
㉘ 有活性的,被激活的
㉙ 卖主
㉚ 促使
㉛ 封装
㉜ 对齐

uses a set of protocols to send its data down the layers, being further encapsulated at each level.^{注4}

This may be **illustrated**③ by an example network **scenario**④, in which two Internet host computers communicate across local network boundaries **constituted**⑤ by their internetworking gateways (routers) shown in Figure 7-1.

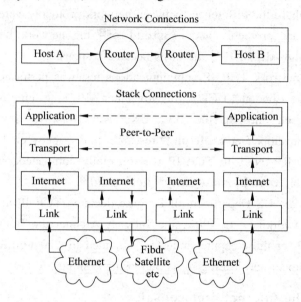

Figure 7-1 TCP/IP stack operating on two hosts connected via two routers and the corresponding layers used at each hop

The functional groups of protocols and methods are the Application Layer, **the Transport Layer**⑥, **the Internet Layer**⑦, **and the Link Layer**⑧. This model was not intended to be a **rigid**⑨ reference model into which new protocols have to fit **in order to**⑩ be accepted as a standard.^{注5} The encapsulation of application data descending through the protocol stack is shown in Figure 7-2.

The following Table 7-1 provides some examples of the protocols grouped in their respective layers.

---

③ 说明,阐明
④ 情景
⑤ 组成
⑥ 传输层
⑦ 网络层
⑧ 链路层
⑨ 严格的
⑩ 为了

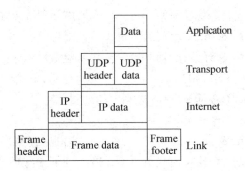

Figure 7-2 Encapsulation of application data descending through the protocol stack

Table 7-1 Some examples of the protocols in their respective layers

Application	DNS, TFTP, TLS/SSL, FTP, Gopher, HTTP, IMAP, IRC, NNTP, POP3, SIP, SMTP, SMPP, SNMP, SSH, Telnet, Echo, RTP, PNRP, rlogin, ENRP
	Routing protocols like BGP and RIP which run over TCP/UDP, may also be considered part of the Internet Layer
Transport	TCP, UDP, DCCP, SCTP, IL, RUDP, RSVP
	OSPF for IPv4 was initially considered IP layer protocol since it runs per IP-subnet, but has been placed on the Link since RFC 2740
Link	ARP, RARP, OSPF (IPv4/IPv6), IS-IS, NDP

3. Implementations

Most computer operating systems in use today, including all **consumer-targeted**❶ systems, include a TCP/IP **implementation**❷.注6 Minimally acceptable implementation includes implementation for (from most **essential**❸ to the less essential) IP, ARP, ICMP, UDP, TCP and sometime IGMP.

Most of the IP implementations are accessible to the programmers using **socket**❹ abstraction (usable also with other protocols) and proper API for most of the operations. This interface is known as BSD sockets and was used **initially**❺ in C.注7

Words

abstract   *adj.*（形容词）   抽象的

---

❶ 消费对象
❷ 执行
❸ 要点
❹ 套接字
❺ 最初

activated	adj.（形容词）	有活性的,被激活的
align	vt. & vi.（动词）	对齐
constitute	vt. & vi.（动词）	组成
construct	n.（名词）	构想,概念
encapaulation	n.（名词）	封装
essential	n.（名词）	要点
evolve	vt. & vi.（动词）	出现；到来
existing	adj.（形容词）；v.（动词）	目前的；现存的；存在(exist 的现在分词）
fundamental	adj.（形容词）	基本的
generation	n.（名词）	产生；一代
hidden	adj.（形容词）	难以发现的,隐藏
illustrate	vt. & vi.（动词）	说明,阐明
initially	adv.（形容词）	最初
interconnection	n.（名词）	互联
logically	adv.（副词）	逻辑上
migration	n.（名词）	迁移
multiple	adj.（形容词）	多重的
operational	adj.（形容词）	可操作的,运行的
permanently	adv.（副词）	永存地,不变地
physically	adv.（副词）	身体上,身体上地,物理地身体上
promote	vt. & vi.（动词）	促进
protocol	n.（名词）	协议
prototype	n.（名词）	原型
reformulation	n.（名词）	再形成；重塑
reliability	n.（名词）	可靠性
rigid	adj.（形容词）	严格的
scenario	n.（名词）	情景
socket	n.（名词）	插座；窝,穴；牙槽
standard	n.（名词）	标准
synthesis	n.（名词）	综合
translate	vt. & vi.（动词）	翻译,转换
transmission	n.（名词）	传输
vendor	n.（名词）	卖主

Phrases

a set of	一套；一组；一副

consumer-targeted	消费对象
in order to	为了
instead of	代替；不是……而是……
open-architecture	开放式体系结构
the internet layer	网络层
the link layer	链路层
the transport layer	传输层

Exercises

【Ex1】 **Answer the questions according to the text：**

(1) What is the Internet Protocol Suite?

(2) What are the most important protocols of Internet Protocol Suite?

(3) What does the TCP/IP model consist of?

(4) When did the Internet Protocol Suite result from research and development conducted by DARPA?

(5) How does an application send its data down the layers?

【Ex2】 **Translate into Chinese：**

(1) The Internet Protocol Suite is the set of communications protocols used for the Internet and other similar networks.

(2) Upper layers are logically closer to the user and deal with more abstract data, relying on lower layer protocols to translate data into forms that can eventually be physically transmitted.

(3) The Internet Protocol Suite resulted from research and development conducted by the Defense Advanced Research Projects Agency (DARPA) in the early 1970s.

(4) Several other TCP/IP prototypes were developed at multiple research centers between 1978 and 1983.

(5) The TCP/IP suite uses encapsulation to provide abstraction of protocols and services.

(6) This model was not intended to be a rigid reference model into which new protocols have to fit in order to be accepted as a standard.

(7) Most computer operating systems in use today, including all consumer-targeted systems, include a TCP/IP implementation.

(8) Most of the IP implementations are accessible to the programmers using socket abstraction (usable also with other protocols) and proper API for most of the operations.

【Ex3】 Choose the best answer：

（1）It is commonly also known as TCP/IP, named from two of the most important protocols in it: the Transmission Control Protocol and _____.

  A. an identifier   B. the Internet Protocol（IP）
  C. an array   D. ITP

（2）The TCP/IP model consists of _____ layers.

  A. three   B. six
  C. four   D. two

（3）The functional groups of protocols and methods are _____, the Transport Layer, the Internet Layer, and the Link Layer.

  A. ActiveX   B. XML
  C. HTML Layer   D. the Application Layer

（4）Most computer operating systems in use today, including all consumer-targeted systems, include a TCP/IP _____.

  A. plan   B. requirement
  C. implementation   D. design

（5）This passage is mainly talking about _____.

  A. Internet Protocol Suite   B. a control
  C. pipelining   D. producing

批 注

注1 Mordern IP networking 是主语；represents 是谓语；several develepments 是宾语；that 引导了一个宾语从句；began 是从句谓语；namely 修饰 Mordern IP networking；Which 引导一个主语从句 emerged 是谓语；together with 修饰主语；Mordern IP networking 属于宾语。

注2 这句话的含义是：1985年,因特网架构理事会举行了一个为期3天由250家厂商代表参加的关于计算产业使用TCP/IP的工作会议,帮助推广协议并且引领它日渐增长的商业应用。

注3 这句话的含义是：这些封包通常与被分离的协议层的一般功能对齐。

注4 这句话的含义是：一般来说,应用层使用一组协议发送数据到下层,这样能更进一步地在每一层中封装数据。

注5 这段话的含义是：这些实用的协议组和方法是应用层、传输层、网际层和网络接口层。这个模型并不是一个严格的参考模型,它并没有要求新协议必须接受它为新的标准。

注6 这句话的含义是：在当今使用的大部分操作系统中,包括所有的用户系统,都含有TCP/IP实现。

注7 这句话的含义是：多数IP操作都能通过程序员使用抽象套接字及合适的API实现大部分的操作。这个接口就是BSD套接,并且它最初使用于C语言中。

Text B Cloud Computing

1. What is cloud computing

Cloud Computing is a resource delivery and **usage**❶ model; it means get resource (Hardware, software) via network. The network of providing resource is called 'Cloud'. The hardware resource in the 'Cloud' seems **scalable**❷ infinitely and can be used whenever.

Cloud Computing refers to both the applications delivered as services over the Internet and the hardware and systems software in the datacenters that provide those services. The services themselves have long been referred to as Software as a Service (SaaS), so we use that **term**❸. The datacenter hardware and software is what we will call a Cloud.

2. New application opportunities

(1) Mobile interactive applications. Tim O'Reilly believes that "the future belongs to services that respond in real time to information provided either by their users or by **nonhuman**❹ sensors." Such services will be attracted to the cloud not only because they must be highly **available**❺, but also because these services generally rely on large data sets that are most conveniently hosted in large datacenters.注1 This is especially the case for services that **combine**❻ two or more data sources or other services, e.g., mash-ups. While not all mobile devices enjoy **connectivity**❼ to the cloud 100% of the time, the challenge of **disconnected**❽ operation has been addressed successfully in specific application **domains**❾, so we do not see this as a significant **obstacle**❿ to **the appeal of**⓫ mobile applications.注2

(2) Parallel batch processing. Although thus far we have **concentrated**⓬ on using Cloud Computing for interactive SaaS, Cloud Computing presents a unique opportunity for

---

❶ 使用、用法
❷ 可伸缩的
❸ 专门名词,术语
❹ 非人类的
❺ 可用的;可获得的
❻ 结合
❼ 连通性
❽ 不连贯的
❾ 区域;畴
❿ 障碍
⓫ 呼吁
⓬ 集中,聚集

batch-processing and **analytics**⑬ jobs that analyze **terabytes**⑭ of data and can take hours to finish. If there is enough data **parallelism**⑮ in the application, users can take advantage of the cloud's new "cost associativity": using hundreds of computers for a short time costs the same as using a few computers for a long time. For example, Peter Harkins, a Senior Engineer at The Washington Post, used 200 EC2 instances (1407 server hours) to **convert**⑯ 17,481 pages of Hillary Clinton's travel documents into a form more friendly to use on the WWW within nine hours after they were **released**⑰. Programming **abstractions**⑱ such as Google's Map Reduce and its **open-source**⑲ counterpart Hadoop allow programmers to express such tasks while hiding the operational **complexity**⑳ of **choreographing**㉑ parallel execution across hundreds of Cloud Computing servers. Indeed, Cloudera is pursuing commercial opportunities in this space. Again, using Gray's insight, the cost/benefit analysis must weigh the cost of moving large datasets into the cloud against the benefit of potential speedup in the data analysis. When we return to economic models later, we **speculate**㉒ that part of Amazon's motivation to host large public datasets for free may be to **mitigate**㉓ the cost side of this analysis and thereby attract users to purchase Cloud Computing cycles near this data.

(3) The rise of analytics. A special case of compute-intensive batch processing is business analytics. While the large database industry was originally dominated by transaction processing, that demand is **leveling off**㉔. A growing share of computing resources is now spent on understanding customers, supply chains, buying habits, ranking, and so on. Hence, while online transaction volumes will continue to grow slowly, decision support is growing rapidly, **shifting**㉕ the resource balance in database processing from transactions to business analytics.

(4) Extension of compute-intensive desktop applications. The latest versions of the **mathematics**㉖ software packages MATLAB and Mathematica are capable of using Cloud Computing to perform expensive evaluations. 注3 Other desktop applications might similarly

---

⑬ 分析学
⑭ 太字节
⑮ 平行度
⑯ 转变;改变
⑰ 发布;释放
⑱ 抽象
⑲ 开源的
⑳ 复杂度
㉑ 筹划,编舞
㉒ 猜测,推测
㉓ 使缓和;减轻
㉔ 趋向平稳;趋平;达到平衡;稳定;平整
㉕ 转移;移动;偏移
㉖ 数学

benet from **seamless**㉗ extension into the cloud. Again, a reasonable test is comparing the cost of computing in the Cloud plus the cost of moving data in and out of the Cloud to the time savings from using the Cloud.注4 **Symbolic mathematics**㉘ involves a great deal of computing per unit of data, making it a domain worth investigating. An interesting alternative model might be to keep the data in the cloud and rely on having **sufficient**㉙ bandwidth to enable suitable **visualization**㉚ and a responsive GUI back to the human user. Offline image **rendering**㉛ or 3D animation might be a similar example: given a compact description of the objects in a 3D scene and the characteristics of the lighting sources, rendering the image is an **embarrassingly**㉜ **parallel**㉝ task with a high computation-to-bytes ratio.

(5) "Earthbound" applications. Some applications that would otherwise be good candidates for the cloud's **elasticity**㉞ and parallelism may be **thwarted**㉟ by data movement costs, the fundamental **latency**㊱ limits of getting into and out of the cloud, or both. For example, while the analytics associated with making long-term financial decisions are appropriate for the Cloud, stock trading that requires microsecond precision is not. Until the cost (and possibly latency) of widearea data transfer decrease (see Section 7), such applications may be less obvious candidates for the cloud.

i. Cloud Computing Economics

In deciding whether hosting a service in the cloud makes sense over the long term, we argue that the **fine-grained**㊲ economic models enabled by Cloud Computing make tradeoff decisions more fluid, and in particular the elasticity offered by clouds serves to transfer risk.

As well, although hardware resource costs continue to decline, they do so at variable rates; for example, computing and storage costs are falling faster than WAN costs. Cloud computing can track these changes-and potentially pass them through to the customer-more effectively than building one's own datacenter, resulting in a closer match of **expenditure**㊳ to actual resource usage.

In making the decision㊴ about whether to move an existing service to the cloud, one must additionally examine the expected average and peak resource utilization, especially if the

㉗ 准确无误的；无缝的
㉘ 符号数学
㉙ 充足的，足够的
㉚ 足够的，充足的
㉛ 渲染
㉜ 令人尴尬地，使人难堪地
㉝ 并行
㉞ 灵活性；伸缩性
㉟ 阻挠
㊱ 潜伏物；潜在因素
㊲ 细粒度
㊳ 花费；支出
㊴ 决策，决定

application may have highly variable **spikes**⑩ in resource demand; the practical limits on real-world utilization of purchased equipment; and various operational costs that vary depending on the type of cloud environment being considered.注5

ii. About The Clouds of Tomorrow

The long dreamed vision of computing as a utility is finally emerging. The elasticity of a utility matches the need of businesses providing services directly to customers over the Internet, as workloads can grow (and shrink) far faster than 20 years ago. It used to take years to grow a business to several million customers-now it can happen in months.

Some question whether companies accustomed to **high-margin**⑪ businesses, such as ad revenue from search engines and traditional packaged software, can compete in Cloud Computing. First, the question presumes that Cloud Computing is a small margin business based on its low cost. Given the typical utilization of medium-sized datacenters, the potential factors of 5 to 7 in economies of scale, and the further savings in **selection**⑫ of cloud datacenter locations, the apparently low costs offered to cloud users may still be highly profitable to cloud providers.注6 Second, these companies may already have the datacenter, networking, and software **infrastructure**⑬ in place for their mainline businesses, so Cloud Computing represents the opportunity for more income at little extra cost.

Words

abstraction	*n.*（名词）	抽象
analytics	*n.*（名词）	分析学
available	*adj.*（形容词）	可用的；可获得的
choreograph	*vt. & vi.*（动词）	筹划，编舞
combine	*vt. & vi.*（动词）	结合
complexity	*n.*（名词）	复杂度
concentrate	*vt. & vi.*（动词）	集中，聚集
connectivity	*n.*（名词）	连通性
convert	*vt. & vi.*（动词）	转变；改变
disconnected	*adj.*（形容词）	不连贯的
domain	*n.*（名词）	区域；畴
elasticity	*n.*（名词）	灵活性；伸缩性
embarrassingly	*adv.*（副词）	令人尴尬地，使人难堪地

---

⑩ 尖状物
⑪ 高利润的
⑫ 选择
⑬ 基础结构，基础设施

expenditure	*n.*（名词）	花费；支出
fine-grain	*adj.*（形容词）	细粒度的
high-margin	*adj.*（形容词）	高利润的
infrastructure	*n.*（名词）	基础结构，基础设施
latency	*n.*（名词）	潜伏物；潜在因素
mathematics	*n.*（名词）	数学
mitigate	*vt. & vi.*（动词）	使缓和；减轻
nonhuman	*adj.*（形容词）	非人类的
obstacle	*n.*（名词）	障碍
parallelism	*n.*（名词）	平行度
release	*vt. & vi.*（动词）	发布；释放
rendering	*n.*（名词）	渲染
scalable	*adj.*（形容词）	可攀登的；可升级的
seamless	*adj.*（形容词）	准确无误的；无缝的
selection	*n.*（名词）	选择
shift	*vt. & vi.*（动词）	转移；移动；偏移
speculate	*vt. & vi.*（动词）	猜测，推测
spike	*n.*（名词）	尖状物
sufficient	*adj.*（形容词）	充足的，足够的
terabyte	*n.*（名词）	兆兆字节，太字节
term	*n.*（名词）	专门名词，术语
thwart	*vt. & vi.*（动词）	阻挠
usage	*n.*（名词）	使用、用法
visualization	*n.*（名词）	可视化

Phrases

in making the decision	在做决定
leveling off	弄平
open-source	开源的
symbolic mathematics	符号数学
the appeal of	上诉的

批 注

注1 这句话的含义是：这种服务是非常适合云的，因为它们不仅要求高可用性，而且通常需要大型数据中心妥善存储大量数据。

注2 这句话的含义是：虽然不是所有的移动设备都能保证一直与云设备连接，但是脱机状态下的

处理在具体应用领域中已经讨论过，所以这里不认为这是影响移动应用的重大障碍。

注3　这句话的含义是：最新版本的数学软件包 MATLAB 和 Mathematica 可以通过云计算进行复杂的评估计算。

注4　重申：一个合理的测试是比较利用云计算的成本，以及数据传输的费用与利用云计算节省的时间。

注5　在决定是否将现有服务转向云计算时，必须仔细考虑预期的平均资源使用和峰值资源使用情况，特别是在应用具有对资源需求高可变性情况下，必须考虑购买设备的实际使用情况以及在不同云环境下的运行操作成本。

注6　Given 分词短语为独立结构成分；may still be 为谓语。这句话的含义是：只要考虑一个中等规模数据中心的利用率，考虑将其扩大5~7倍，并考虑未来对云计算数据中心地点的选择，虽然云用户支付的费用不高，但最终云服务提供商仍可获得可观的利润。

Associated Reading

Edraw Network Diagram v5.6

EDraw Network Diagrammer is a professional network diagramming software with rich examples and templates. Easy to draw detailed physical, logical, Cisco and network architecture diagrams, using a comprehensive set of network and computer equipment shapes shown as Figure 7-3.

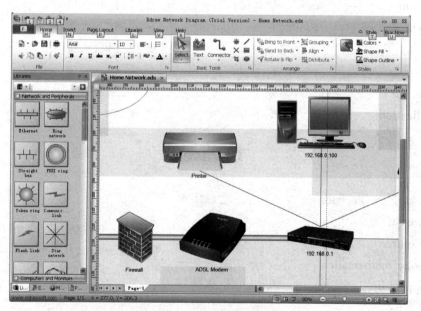

Figure 7-3　Edraw network diagram v5.6

Pre-drawn network diagram icons representing computers, network devices plus smart connectors help design diagram network, create accurate network diagrams and documentation to be used in your network diagram project. Abundant network diagram templates, network diagram symbols and network diagram examples will help to quickly create most common network diagrams. Intuitive interface helps to create accurate diagrams in a minutes. Just drag and drop pre-drawn shapes representing computers and network devices. Double click and set equipment data. Create detailed physical, logical and network architecture diagrams, using a comprehensive set of network and computer equipment shapes.

Unit 8　Data Structure

Text A　Data Structures❶ and Algorithms❷ in Java

Once you've learned to program, you run into real-world problems that require more than a programming language alone to solve.^{注1} Data Structures and Algorithms in Java is a gentle **immersion**❸ into the most **practical**❹ ways to make data do what you want it to do.^{注2}

1. Overview of data structures

Another way to look at data structures is to focus on their strengths and weaknesses. In this section we'll provide an overview, **in the form of**❺ a table as Table 8-1, of the major data storage structures we'll be discussing.

Table 8-1　Characteristics❻ of Data Structures

Data Structure	Advantages	Disadvantages
Array	Quick insertion, very fast access if index known	Slow search, slow deletion, fixedsize^{注3}
Ordered array	Quicker search than unsorted array	Slow insertion and deletion, fixed size
Stack	Provides last-in, first-out access	Slow access to other items
Linked list	Quick insertion, quick deletion	Slow search
Binary tree	Quick search, insertion, deletion (if tree remains balanced)	Deletion algorithm is complex^{注4}
Red-black tree	Quick search, insertion, deletion. Tree always balanced	Complex
Hash table	Very fast access if key known. Fast insertion	Slow deletion, access slow if key not known, inefficient memory usage^{注5}
Graph	Models real-world situations	Some algorithms are slow and complex^{注6}

---

❶ 结构体
❷ 算法
❸ 沉浸;浸没;浸
❹ 实用的
❺ 以……形式
❻ 特性,特点

2. Overview of algorithms

Many of the algorithms we'll discuss apply directly to specific data structures. For most data structures, you need to know how to

(1) Insert a new data item.

(2) Search for a specified item.

(3) Delete a specified item.

You may also need to know how to iterate through all the items in a data structure, visiting each one in turn so as to display it or perform some other action on it. One important algorithm **category** ❼ is sorting. There are many ways to sort data, The concept of recursion is important in designing certain algorithms.

3. Object-Oriented❽ programming

OOP was invented because **Procedural**❾ languages, such as C, Pascal, and BASIC, were found to **be inadequate**❿ **for** large and complex programs. Why was this?

The problems have to do with the overall organization of the program. <u>Procedural programs are organized by dividing the code into functions (called procedures or subroutines in some languages).</u> 注7 Groups of functions could form larger units called modules or files. Crude Organizational Units One difficulty with this kind of function-based organization was that it focused on functions at the expense of data. There weren't many options when it came to data. To simplify slightly, data could be local to a particular function or it could be global—**accessible**⓫ to all functions. There was no way (at least not a flexible way) to specify that some functions could access a **variable**⓬ and others couldn't. This caused problems when several functions needed to access the same data. <u>To be available to more than one function, such variables had to be global, but global data could be accessed **inadvertently**⓭ by any function in the program.</u> 注8 This leads to frequent programming errors. What was needed was a way to fine-tune data accessibility, allow in variables to be available to functions with a need to access it, but hiding it from others.

4. Objects in a nutshell

From the idea of objects arose in the programming community as a solution to the

---

❼ 分类,类别

❽ 面向对象的

❾ 过程的

❿ 不满意的,不足的

⓫ 可以访问的

⓬ 变量

⓭ 不经意的,随意的

problems with procedural languages.

(1) Objects

This new **entity**⑭, the object, solves several problems **simultaneously**⑮. Not only does a programming object correspond more **accurately**⑯ to objects in the real world, it also solves the problem **engendered by**⑰ global data in the procedural model.注9

(2) Classes

You might think that the idea of an object would be enough for one programming **revolution**⑱, but there's more. 注10 Early on, it was realized that you might want to make several objects of the same type. Maybe you're writing a **furnace control program**⑲ for an entire apartment house, for example, and you need several dozen **thermostat objects**⑳ in your program. It seems a shame to go to the trouble of specifying each one **separately**㉑. Thus, the idea of classes was born.

Words

accessible	*adj.*（形容词）	可以访问的
accurately	*adv.*（副词）	准确地
category	*n.*（名词）	分类,类别
characteristic	*n.*（名词）	特性,特点
entity	*n.*（名词）	实体,实数
immersion	*n.*（名词）	沉浸;浸没;浸
inadequate	*adj.*（形容词）	不满意的,不足的
inadvertently	*adv.*（副词）	不经意
orient	*adj.*（形容词）	面向……的,朝向
practical	*adj.*（形容词）	实用的
procedural	*adj.*（形容词）	过程的
revolution	*n.*（名词）	革命
separately	*adv.*（副词）	分别地
simultaneously	*adv.*（副词）	同时地
structure	*n.*（名词）	结构体

⑭ 实体,实数
⑮ 同时的
⑯ 准确的
⑰ 由……产生的
⑱ 革命
⑲ 温控的程序
⑳ 温控器对象
㉑ 分开的

| variable | *n.* (名词) | [数]变量 |

Phrases

be inadequate for	不满意的,不足的
engendered by	由……产生
furnace control program	温控的程序
in the form of	以……的形式
thermostat objects	温控器对象

Exercises

【Ex1】 **Answer the questions according to the text:**

(1) What are the advantages and disadvantages of Array?
(2) What are the advantages and disadvantages of Binary Tree?
(3) What is algorithm?
(4) What's the problem with Procedural Languages?
(5) What are Object and Class?

【Ex2】 **Translate into Chinese:**

(1) Data Structures and Algorithms in Java is a gentle immersion into the most practical ways to make data do what you want it to do.
(2) The concept of recursion is important in designing certain algorithms.
(3) Groups of functions could form larger units called modules or files.
(4) Not only does a programming object correspond more accurately to objects in the real world, it also solves the problem engendered by global data in the procedural model.
(5) Global data could be accessed inadvertently by any function in the program.
(6) Linked lists are probably the second most commonly used general-purpose storage structures after arrays.
(7) The linked list is a versatile mechanism suitable for use in many kinds of general-purpose databases.
(8) For some kinds of hash tables, performance may degrade catastrophically when the table becomes too full.

【Ex3】 **Choose the best answer:**

(1) _____ in Java is a gentle immersion into the most practical ways to make data do what you want it to do.

    A. Data analysis   B. Requirement
    C. Data Structures and Algorithms   D. Design

（2）Which is the Array's advantage _____.
   A. Provides last-in, first-out access
   B. Quick insertion, very fast access if index known
   C. Quicker search than unsorted array
   D. Models real-world situations

（3）OOP was invented because procedural languages, such as _____, Pascal, and BASIC.
   A. C   B. C++   C. C#   D. ASP

（4）The _____ is an entity to solve several problems simultaneously.
   A. Object   B. Class   C. Method   D. None

（5）_____ are organized by dividing the code into functions.
   A. Parallel programs   B. Procedural programs
   C. Structure programs   D. None

批 注

注1　这句话的含义是：当你学习编程的时候，你遇到的现实问题需要不只一种编程语言才可以得到解决。

注2　这句话的含义是：基于Java的数据结构与算法是把数据按你想要的方式存储的最实用的方式。

注3　这行的含义是：数组的优点是插入快，如果知道索引,可以非常快速地存取。缺点是查找慢,删除慢,定长。

注4　这行的含义是：二叉树的优点是查找、插入及删除(如果树保持了平衡)快速。缺点是删除算法很复杂。

注5　这行的含义是：哈希表的优点是如果知道主键,访问很快,快速地插入。缺点是删除慢,如果不知道主键,那么内存的利用率将很低。

注6　这行的含义是：图的优点是对现实的情形进行建模。缺点是有些算法很慢,很复杂。

注7　这句话的含义是：进程程序把代码按功能进行划分。

注8　global data 为主语；by any function in the program 为方式状语；To be available to more than one function 表示目的；这句话的含义是：为了获得更多的功能，有些变量就设置为全局变量，但是全局变量可以被程序中的任何一个函数访问。

注9　这句话的含义是：对象编程方法不仅更精确地对象对应于现实世界中的对象，而且还解决了过程模型中全局数据产生的问题。

注10　这句话的含义是：你可能认为对象的概念对编程革命已经足够了，但实际上还不够。

Text B The Introduction of Two Important Data Structures

In this part we will take a brief introduction about two important data structures, Linked list and Hash table, which will make it effective to search or store the data.

1. Linked❶ list

We saw that arrays had certain disadvantages as data storage structures. In an unordered array, searching is slow, whereas in an ordered array, insertion is slow. 注1 In both kinds of arrays deletion is slow. Also, the size of an array can't be changed after it's created. In this part we'll look at a data storage structure that solves some of these problems: the linked list.

Linked lists are probably the second most commonly used general-purpose storage structures after arrays. The linked list is a **versatile**❷ **mechanism**❸ suitable for use in many kinds of general-purpose databases. It can also replace an array as the basis for other storage structures such as stacks and queues. In fact, you can use a linked list in many cases where you use an array (unless you need frequent **random access**❹ to individual items using an index). Linked lists aren't the solution to all data storage problems, but they are surprisingly versatile and **conceptually**❺ simpler than some other popular structures such as trees. 注2 We'll **investigate**❻ their strengths and weaknesses as we go along. In this section we will refer to two very useful data structure.

2. Links

In a linked list, each data item is embedded in a link. A link is an object of a class called something like Link. Because there are many similar links in a list, it makes sense to use a separate class for them, **distinct**❼ from the linked list itself. 注3 Each link objects contains a reference (usually called next) to the next link in the list. A field in the list itself contains are reference to the first link. Here's part of the definition of a class Link. It contains some data and a reference to the next link.

```
class Link
{
public int iData;              //data
public double dData;           //data
public Link next;              //reference to next link
    }
```

This kind of class definition is sometimes called self-referential because it contains a field—called next in this case—of the same type as itself. 注4

---

❶ 链接的
❷ 易变的,灵活的
❸ 机制
❹ 随机存取
❺ 概念上的
❻ 探讨,投入
❼ 不同的

3. Hash[8] tables

A hash table is a data structure that offers very fast insertion and searching. When you first hear about them, hash tables sound almost too good to be true. No matter how many data items there are, insertion and searching (and sometimes deletion) can **take close to**[9] constant time: O(1) in Big O notation. 注5 In practice this is just a few machine instructions. For a human user of a hash table this is **essentially**[10] instantaneous. It's so fast that computer programs typically use hash tables when they need to look up tens of thousands of items in less than a second (as in spelling checkers). Hash tables are **significantly**[11] faster than trees Not only are they fast, hash tables are **relatively**[12] easy to program. Hash tables do have several disadvantages. They're based on arrays, and arrays are difficult to expand once they've been created. For some kinds of hash tables, **performance**[13] may **degrade**[14] **catastrophically**[15] when the table becomes too full, so the programmer needs to have a fairly accurate idea of how many data items will need to be stored (or be prepared to **periodically**[16] transfer data to a larger hash table, a **time-consuming**[17] process). 注6

Also, there's no convenient way to visit the items in a hash table in any kind of order (such as from smallest to largest). If you need this capability, you'll need to look elsewhere. However, if you don't need to visit items in order, and you can predict in advance the size of your database, hash tables are **unparalleled in**[18] speed and convenience. 注7

Words

catastrophically	*adv.* （副词）	严重地
conceptually	*adv.* （副词）	概念上地
degrade	*vt. & vi.* （动词）	下降
distinct	*adj.* （形容词）	不同的
essentially	*adv.* （副词）	本质上

- [8] 哈希
- [9] 接近；趋近
- [10] 本质上
- [11] 相当的，有意义的
- [12] 相对的
- [13] 执行，性能
- [14] 下降
- [15] 严重地
- [16] 阶段性地
- [17] 耗时的
- [18] 不可匹配的

hash	*n.*（名词）	哈希
investigate	*vt. & vi.*（动词）	探讨,投入
linked	*adj.*（形容词）	连接的
mechanism	*n.*（名词）	机制
performance	*n.*（名词）	执行,性能
periodically	*adv.*（副词）	阶段性地
relatively	*adv.*（副词）	相对地
significantly	*adv.*（副词）	相当地,有意义地
time-consuming	*adj.*（形容词）	耗时的
versatile	*adj.*（形容词）	易变的,灵活的

Phrases

random access	随机存取
take close to	接近
unparalleled in	不可匹配的
round-robin	时间片轮转
blocks of	许多块
scattered throughout	散布在

批 注

注1　这句话的含义是：在无排序的数组中,数据查找速度较慢,而在有序数组中,数据插入操作速度较慢。

注2　这句话的含义是：链接表并不能解决所有的数据存储问题,但是它们比其他一些常见结构,如树的功能更强,概念上也更简单。

注3　这句话的含义是：由于在一个表里有很多相似的链接,因此用一个分开的类区别它们很有必要。

注4　这句话的含义是：这种类的定义有时称为自引用,因为它包含一个它自己的类定义。

注5　No matter how 为状语从句；insertion and searching（and sometimes deletion）为句子的主语；can take close to 为谓语。这句话的含义是：不管有多少个数据项,插入及查找所花时间差不多,即O(1)的时间级。

注6　这句话的含义是：对于一些类型的哈希表,当表较满的时候,执行的效率将大大降级,因此程序员必须清楚地知晓有多少数据项将被保存(或将数据阶段性地传输到大的哈希表,这比较消耗时间)。

注7　这句话的含义是：然后,如果你不需要有序地访问数据项,也可以预先预计数据库的大小,那么哈希表可能在速度和效率方面不匹配。

Associated Reading

Lists

The array implementation of our collection has one serious drawback: you must know the maximum number of items in your collection when you create it. This presents problems in programs in which this maximum number cannot be predicted accurately when the program starts up. Fortunately, we can use a structure called a linked list to overcome this limitation.

1. Circularly linked lists

By ensuring that the tail of the list is always pointing to the head, we can build a circularly linked list.<u>注1</u> If the external pointer (the one in struct t_node in our implementation), points to the current "tail" of the list, then the "head" is found trivially via tail->next, permitting us to have either LIFO or FIFO lists with only one external pointer shown as Figure 8-1. In modern processors, the few bytes of memory saved in this way would probably not be regarded as significant. A circularly linked list would more likely be used in an application which required "round-robin" scheduling or processing.<u>注2</u>

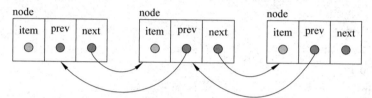

Figure 8-1　Circularly linked lists

Doubly linked lists have a pointer to the preceding item as well as one to the next.

They permit scanning or searching of the list in both directions. (To go backwards in a simple list, it is necessary to go back to the start and scan forwards.) Many applications require searching backwards and forwards through sections of a list: for example, searching for a common name like "Kim" in a Korean telephone directory would probably need much scanning backwards and forwards through a small region of the whole list, so the backward links become very useful.<u>注3</u> In this case, the node structure is altered to have two links:

```
struct t_node {
void * item;
struct t_node * previous;
struct t_node * next;
} node;
```

2. Lists in arrays

Although this might seem pointless (Why impose a structure which has the overhead of the "next" pointers on an array?), this is just what memory allocators do to manage available space.

Memory is just an array of words.<u>注4</u> After a series of memory allocations and deallocations, there are blocks of[19] free memory scattered[20] throughout the available heap space. In order to be able to re-use this memory, memory allocators will usually link freed blocks together in a free list by writing pointers to the next free block in the block itself.[注5] An external free list pointer points to the first block in the free list. When a new block of memory is requested, the allocator will generally scan the free list looking for a freed block of suitable size and delete it from the free list (re-linking the free list around the deleted block).[注6] Many variations of memory allocators have been proposed: refer to a text on operating systems or implementation of functional languages for more details. The entry in the index under garbage collection will probably lead to a discussion of this topic.

批 注

注1 这句话的含义是：如果要保证一个表尾总是指向表头，须构建一个循环链表。

注2 这句话的含义是：一个循环链表将在这种情况下应用，即需要时间片轮转调度或处理。

注3 这句话的含义是：许多的应用程序需要通过列表部分进行向后或向前搜索。例如，在韩语电话簿目录中，要查找一个名为 Kim 的常见名字，可能需要通过对小部分的电话簿进行向前或向后扫描，因此向后链表就变得很有用了。

注4 这是一个简单句，这句话的含义是：内存就像是一组字的数组。

注5 In order to be able to re-use this memory 作目的状语；memory allocators 是句子的主语；by writing pointers to the next free block in the block itself 是方式状语。这句话的含义是：为了能够对内存进行重利用，内存分配器将通过指针把空闲的块链接起来。

注6 When a new block of memory is requested, 引导一个状语从句；looking for a freed block of suitable size 作伴随状语；and delete it from the free list 与 scan 并列。

[19] 许多块
[20] 分发

Unit 9　Microsoft Developer Network

Text A　What is MSDN?

The Microsoft Developer Network (MSDN) is the portion of Microsoft responsible for managing the firm's relationship with developers and testers: hardware developers interested in the operating system (OS), developers standing on the various OS **platforms**❶, developers leveraging the API and scripting languages of Microsoft's many applications.^{注1} The relationship management is situated in **assorted**❷ media: web sites, newsletters, developer **conferences**❸, trade media, blogs and DVD **distribution**❹. The life cycle of the relationships ranges from **legacy**❺ support through **evangelizing potential offerings**❻.

1. History

The service started in 1992, but **initially**❼ only the Microsoft Developer Network CD-ROM was **available**❽. A Level II_**subscription**❾ was added in 1993, that included the MAPI, ODBC, TAPI and VFW SDKs. MSDN2 was opened in November 2004 as a source for Visual Studio 2005 API information, with **noteworthy**❿ differences being updated web site code, **conforming**⓫ better to web standards and thus giving a long awaited improved support for alternative web browsers to Internet Explorer in the API browser. In 2008, the original MSDN **cluster**⓬ was retired and MSDN2 became msdn.microsoft.com.

In 1992, Bob Gunderson began writing a column in the MSDN Developer News (an actual paper-based publication) using the **pseudonym**⓭ "Dr. GUI". The column provided

---

❶ 平台
❷ 把……分类
❸ 会议
❹ 分发,分配
❺ 遗产
❻ 传福音的潜力产品
❼ 最初地
❽ 可利用的
❾ 预定
❿ 值得注意的
⓫ 遵守,适应,相似,一致,符合
⓬ 串,簇,群,组
⓭ 假名;化名;(尤指)笔名

answers to questions **submitted**[14] by MSDN **subscribers**[15]. The **caricature**[16] of Dr. GUI was based on a photo of Bob. When he left the MSDN team Dennis Crain took over the Dr. GUI role and added medical humor to the column. Upon his **departure**[17], Dr. GUI became the **composite**[18] identity of the original group (most notably Paul Johns) of Developer Technology Engineers that provided in-depth technical articles to the Library. All-in-all, it was a good place to from which to put all MSDN tasks in **perspective**[19]. Ken Lassesen produced the original system (Panda) to publish MSDN on the Internet and in HTML instead of the earlier multimedia viewer engine. Dale Rogerson, Nigel Thompson and Nancy Cluts all published MS Press books while on the MSDN team. As of August 2010, few around Microsoft remember Dr.

2. Information service

The **division**[20] runs an information service provided by Microsoft for software developers. Its main focus is on Microsoft's .NET platform, however it also features articles on areas such as programming practices and design **patterns**[21]. Many resources are available for free online, while others are available by mail **via**[22] a subscription.

Depending on subscription level, **subscribers**[23] may receive early editions of Microsoft operating systems or other Microsoft products (Microsoft Office applications, Visual Studio, etc.).

Universities and high schools can **enroll**[24] in the MSDN Academic Alliance program, which provides access to some Microsoft developer software for their computer science and engineering students (and possibly other students or **faculty**[25] as well). An MSDNAA account is not an MSDN account and cannot be used to access the subscriber's section of the MSDN website or its downloads.

3. Software subscriptions

MSDN has historically offered a subscription **package**[26] whereby developers have access

---

[14] 提交
[15] 捐款人
[16] 讽刺
[17] 离开;出发
[18] 混合成的,综合成的
[19] 视觉,观点,远景
[20] 分,除,部门
[21] 型,模式,花样
[22] 经由,通过
[23] 订阅者
[24] 入学,加入
[25] 教员
[26] 包,包裹

and **licenses**❷ to use nearly all Microsoft software that has ever been released to the public. Subscriptions are sold on an **annual**❷ basis, and cost up to $10,939 USD per year per subscription, as it is offered in several **tiers**❷. Holders of such subscriptions (except the lowest library-only levels) receive new Microsoft software on DVDs or via downloads every few weeks or months.^{注2} The software generally comes on specially marked MSDN discs, but contains the **identical**❸ **retail**❸ or **volume**❷-license software as it is released to the public.

Although in most cases the software itself functions exactly like the full product, the MSDN end-user license agreement **prohibits**❸ use of the software in a business production environment.^{注3} This is a legal **restriction**❸, not a technical one. As an example, MSDN regularly includes the latest Windows operating systems (such as Windows Vista and Windows 7), server software such as SQL Server 2008, development tools such as Visual Studio, and applications like Microsoft Office and MapPoint. For software that requires a product key, a Microsoft website **generates**❺ these on demand. Such a package provides a single computer **enthusiast**❻ with access to nearly everything Microsoft offers. However, a business **caught with**❼ an office full of PCs and servers running the software included in an MSDN subscription without the **appropriate**❽ non-MSDN licenses for those machines would be treated no differently in a software licensing audit than if the software were obtained through **piracy**❾.^{注4}

Words

annual	*adj.* (形容词)	年度的
apporpriate	*adj.* (形容词)	适当的
assort	*vt. & vi.* (动词)	把……分类
available	*adj.* (形容词)	可利用的
caricature	*vt. & vi.* (动词)	讽刺

❷ 许可证,执照,特许
❷ 年度的
❷ 层,等级
❸ 相同的,同一的
❸ 零售
❷ 卷,册,批量
❸ 禁止,不准
❸ 限制,限定,约束
❺ 生成,产生
❻ 热衷者,渴慕者
❼ 被抓住
❽ 适当的
❾ 非法翻印,海盗行为,盗版

cluster	n.（名词）	串，簇，群，组
composite	adj.（形容词）	混合成的，综合成的
conference	n.（名词）	会议
conform	vt. & vi.（动词）	遵守，适应，相似一致，符合
departure	vt. & vi.（动词）	离开，出发
distribution	n.（名词）	分发，分配
division	n.（名词）	分，除，部门
enroll	vt. & vi.（动词）	入学，加入，招收，吸收
enthusiast	n.（名词）	热衷者，渴慕者
faculty	n.（名词）	教员
generate	vt. & vi.（动词）	生成，产生
identical	adj.（形容词）	相同的，同一的
initially	adv.（副词）	最初地
legacy	n.（名词）	遗产
license	n.（名词）	许可证，执照，特许
noteworthy	adj.（形容词）	值得注意的
package	n.（名词）	包，包裹；套装软件，程序包
pattem	n.（名词）	型，模式，花样
perspective	n.（名词）	视觉，观点，远景
piracy	n.（名词）	非法翻印，海盗行为
platform	n.（名词）	平台
prohibit	vt. & vi.（动词）	禁止，不准
pseudonym	n.（名词）	假名；化名；（尤指）笔名
restriction	n.（名词）	限制，限定，约束
retail	n.（名词）	零售
submit	vt. & vi.（动词）	服从，听从
subscriber	n.（名词）	订阅者，认股人；捐款人
subscription	n.（名词）	预订，预约，捐款
tier	n.（名词）	层，等级
via	vt. & vi.（动词）	经由，通过
volume	n.（名词）	卷，册，批量

Phrases

caught with	捉
evangelizing potential offerings	传福音的潜力产品

Exercises

【Ex1】 Fill in the blanks according to the text:

Although in most cases the software itself functions exactly like the full product, the MSDN end-user license agreement _____ use of the software in a business production environment. This is a legal _____, not a technical one. As an example, MSDN regularly includes the latest Windows operating systems (such as Windows Vista and Windows 7), server software such as SQL Server 2008, development tools such as Visual Studio, and applications like Microsoft Office and MapPoint. For software that requires a product key, a Microsoft website _____ these on demand. Such a package provides a single computer _____ with access to nearly everything Microsoft offers. However, a business caught with an office full of PCs and servers running the software included in an MSDN subscription without the _____ non-MSDN licenses for those machines would be treated no differently in a software licensing audit than if the software were obtained through piracy.

【Ex2】 Translate into Chinese:

(1) The Microsoft Developer Network (MSDN) is the portion of Microsoft responsible for managing the firm's relationship with developers and testers: hardware developers interested in the operating system (OS), developers standing on the various OS platforms, developers leveraging the API and scripting languages of Microsoft's many applications.

(2) When he left the MSDN team Dennis Crain took over the Dr. GUI role and added medical humor to the column.

(3) Component is the default implementation of IComponent and serves as the base class for all components in the common language runtime.

(4) In this case, you can derive your own class from the AsyncCompletedEventArgs class and provide additional private instance variables and corresponding read-only public properties.

(5) The System.ComponentModel namespace provides classes that are used to implement the run-time and design-time behavior of components and controls.

(6) An MSDNAA account is not an MSDN account and cannot be used to access the subscriber's section of the MSDN website or its downloads.

(7) All-in-all, it was a good place to from which to put all MSDN tasks in perspective.

(8) If you add an instance of the system.AsyncCompletedEventHandler delegate to the event.

【Ex3】 Choose the best answer:

(1) In this case, you can derive your own class from the AsyncCompletedEventArgs class and provide _____ private instance variables and corresponding read-only public properties.

  A. additional B. appropriate C. alert D. available

(2) Many resources are available for free online, while others are available by mail _____ a subscription.

  A. over B. by C. through D. via

(3) The service started in 1992, but _____ only the Microsoft Developer Network CD-ROM was available.

  A. interact B. initially C. identical D. inherits

(4) MSDN2 was opened in November 2004 as a source for Visual Studio 2005 API information, with noteworthy differences being updated web site code, conforming better _____ web standards and thus giving a long awaited improved support for alternative web browsers to Internet Explorer in the API browser.

  A. out B. up C. on D. to

(5) This namespace includes the base classes and interfaces for _____ attributes and type converters, binding to data sources, and licensing components.

  A. executing B. performing C. implementing D. enforcing

批 注

注1 这句话的含义是：Microsoft 网络开发平台是微软的一部分，用于更好地处理开发和测试人员间的协调关系，硬件开发人员负责的是操作系统(OS)，开发人员面对的是不同的 OS 平台，开发利用该 API 和微软的许多应用程序语言的脚本。

注2 这句话的含义是：这种订阅人(除最低水平外)会有 DVD，或通过每隔数周或数月下载新的微软软件。

注3 Although in most cases the software itself functions exactly like the full product 引导一个状语从句。这句话的含义是：虽然在大多数情况下软件的功能和完整的产品完全一样，但 MSDN 的最终用户许可协议禁止在企业的生产环境中使用软件。

注4 caught with an office full of PCs and servers 为后置定语，修饰 business。

Text B Getting to Know MSDN

MSDN's primary web presence at msdn.microsoft.com is a collection of sites for the developer community that provide information, **documentation**❶, and discussion which is

---

❶ 文件

authored both by Microsoft and by the community at large. Recent **emphasis**❷ on and **incorporation of**❸ applications such as **forums**❹, blogs, library annotations, and social bookmarking are changing the nature of the MSDN site from a one-way information service to an open dialog between Microsoft and the developer community. The main website and most of its **constituent**❺ applications below are available in 56 or more languages.

1. System. ComponentModel namespace

The System. ComponentModel namespace provides classes that are used to implement the run-time and design-time behavior of components and controls.注1 This namespace includes the base classes and interfaces for **implementing**❻ attributes and type converters, binding to data sources, and licensing components.

The classes in this namespace divide into the following **categories**❼:

- Core component classes. See the **Component**, **IComponent**, **Container**, and **IContainer classes**❽.
- Component licensing. See the License, LicenseManager, LicenseProvider, and LicenseProviderAttribute classes.
- Attributes. See the Attribute class.
- Descriptors and **persistence**❾. See the TypeDescriptor, EventDescriptor, and PropertyDescriptor classes.
- Type converters. See the TypeConverter class.

2. AsyncCompletedEventArgs class

It provides data for the MethodNameCompleted event.

If you are using a class that implements the Event-based Asynchronous Pattern Overview, the class will provide a MethodNameCompleted event.注2 If you add an instance of the System. ComponentModel. AsyncCompletedEventHandler delegate to the event, you will receive information about the outcome of a **synchronous**❿ operations in the AsyncCompletedEventArgs parameter of the **corresponding**⓫ event-handler method.

The client application's event-handler delegate can check the Cancelled property to

---

❷ 强调,重点
❸ 纳入
❹ 论坛
❺ 选民,成分,组分
❻ 使生效,履行,实施
❼ 种类,类别
❽ 分别表示:组件类,组件接口类,容器类,容器接口类
❾ 坚持
❿ 同时发生的,同步的
⓫ 相应的,相当的

determine if the asynchronous task was cancelled. 注3

The client application's event-handler delegate can check the Error property to determine if an **exception**⑫ occurred **during execution of the asynchronous task**⑬.

If the class supports multiple asynchronous methods, or multiple calls to the same asynchronous method, you can determine which task raised the MethodNameCompleted event by checking the value of the UserState property. 注4 Your code will need to track these tokens, known as task IDs, as their corresponding asynchronous tasks start and complete.

Classes that follow the Event-based Asynchronous Pattern can raise events to **alert**⑭ clients about the status of **pending**⑮ asynchronous operations. If the class provides a MethodNameCompleted event, you can use the AsyncCompletedEventArgs to tell clients about the outcome of asynchronous operations.

You may want to communicate to clients more information about the outcome of an asynchronous operation than an AsyncCompletedEventArgs accommodates. 注5 In this case, you can derive your own class from the AsyncCompletedEventArgs class and provide **additional**⑯ private **instance**⑰ variables and corresponding read-only public properties. Call the RaiseExceptionIfNecessary method before returning the property value, **in case**⑱ the operation was canceled or an error occurred.

3. IComponent interface

Provides functionality required by all components.

Component is the default implementation of IComponent and serves as the base class for all components in the common language runtime.

You can contain components in a container. In this **context** ⑲, containment refers to logical containment, not visual containment. 注6 You can use components and containers **in a variety of**⑳ **scenarios**㉑, both **visual**㉒ and non visual.

System. Windows. Forms. Control **inherits**㉓ from Component, the default implementation of IComponent. 注7

---

⑫ 例外
⑬ 异步任务执行过程中
⑭ 警觉的
⑮ 直到,等到……期间
⑯ 添加的,额外的
⑰ 实例
⑱ 万一
⑲ 上下文
⑳ 背景
㉑ 情景
㉒ 视觉的,看得见的
㉓ 继承

A component **interacts**㉔ with its container primarily through a container-provided ISite, which is a **repository**㉕ of container-specific per-component information.

To be a component, a class must implement the IComponent interface and provide a basic constructor that requires no **parameters**㉖ or a single parameter of type IContainer. For more information about implementing IComponent, see Programming with Components.

Words

additional	adj.（形容词）	添加的,额外的
alert	adj.（形容词）	警觉的
category	n.（名词）	种类,类别
Component, IComponent, Container, and IContainer classes	n.（专有名词）	组件类,组件接口类,容器类,容器接口类
constituent	n.（名词）	选民,成分,组分
corresponding	adj.（形容词）	相应的,相当的
documentation	n.（名词）	文件
emphasis	n.（名词）	强调,重点
exception	n.（名词）	例外
forums	n.（名词）	论坛
Implement	vt. & vi.（动词）	使生效,履行,实施
inherit	vt. & vi.（动词）	继承
instance	n.（名词）	实例,实体
interact	vt. & vi.（动词）	相互作用,相互影响
parameter	n.（名词）	界限,范围,参数
pending	prep.（介词）	直到,等到……期间
persistence	n.（名词）	坚持
repository	n.（名词）	储藏室,仓库
scenarios	n.（名词）	情景
synchronous	adj.（形容词）	同时发生的,同步的
visual	adj.（形容词）	视觉的,看得见的

Phrases

incorporation of	纳入
in case	万一;假使

㉔ 相互作用,相互影响
㉕ 储藏室,仓库
㉖ 界限,范围,参数

| in a variety of | 在不同的 |

批 注

注1 这句话的含义是：System. ComponentModel 命名空间提供了一些用于实现运行时和设计时组件和控件的行为类。

注2 这句话的含义是：如果您使用的类实现了基于事件的同步模式概述，该类将提供一个 MethodNameCompleted 事件。

注3 这句话的含义是：客户端应用程序的事件处理程序委托可以检查 Cancelled 属性，以确定是否异步任务被取消。

注4 这句话的含义是：如果类支持多个异步方法，或多次调用同一个异步方法，你可以通过检查 UserState 属性决定哪一个任务触发 MethodNameCompleted 事件。

注5 这句话的含义是：你可能想传达给客户更多的异步操作结果，而不只是 AsyncCompletedEventArgs 资料。

注6 这句话的含义是：在这种情况下，容器是指逻辑上的，而不是可视化容器。

注7 这句话的含义是：System. Windows. Forms. Control 从 Component 中继承，默认是 IComponent。

Associated Reading

Types of Bitmaps

A bitmap is an array of bits that specifies the color of each pixel in a rectangular array of pixels. The number of bits devoted to an individual pixel determines the number of colors (Table 9-1) that can be assigned to that pixel shown as Figure 9-1. For example, if each pixel is represented by 4 bits, then a given pixel can be assigned one of 16 different colors ($2^4 = 16$). The following table shows a few examples of the number of colors that can be assigned to a pixel represented by a given number of bits.

Table 9-1 Number of color

Bits per pixel	Number of colors that can be assigned to a pixel
1	$2^1 = 2$
2	$2^2 = 4$
4	$2^4 = 16$
8	$2^8 = 256$
16	$2^{16} = 65\ 536$
24	$2^{24} = 16\ 777\ 216$

Disk files that store bitmaps usually contain one or more information blocks that store information such as number of bits per pixel, number of pixels in each row, and number of rows in the array. Such a file might also contain a color table (sometimes called a color palette). A color table maps numbers in the bitmap to specific colors. The following illustration shows an enlarged image along with its bitmap and color table. Each pixel is represented by a 4-bit number, so there are $2^4 = 16$ colors in the color table. Each color in the table is represented by a 24-bit number: 8 bits for red, 8 bits for green, and 8 bits for blue. The numbers are shown in hexadecimal (base 16) form: A = 10, B = 11, C = 12, D = 13, E = 14, F = 15.

Look at the pixel in row 3, column 5 of the image shown as Figure 9-1. The corresponding number in the bitmap is 1. The color table tells us that 1 represents the color red, so the pixel is red. All the entries in the top row of the bitmap are 3. The color table tells us that 3 represents blue, so all the pixels in the top row of the image are blue.

Note Some bitmaps are stored in bottom-up format; the numbers in the first row of the bitmap correspond to the pixels in the bottom row of the image.

3	3	3	3	3	3	3	3
0	1	4	1	4	1	4	0
0	4	1	4	1	4	1	0
0	5	5	5	5	5	5	0
0	5	5	5	5	5	5	0
0	1	4	1	4	1	4	0
0	4	1	4	1	4	1	0
2	2	2	2	2	2	2	2

0	000000
1	FF0000
2	00FF00
3	0000FF
4	FFFFFF
5	FFFF00
6	FF00FF
7	00FFFF
8	FF0080
9	FF8040
A	804000
D	008080
C	800000
D	800080
E	8080FF

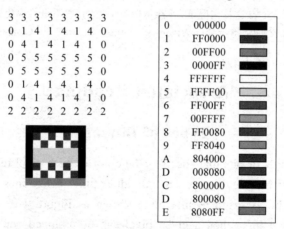

Figure 9-1 A bitmap using color talbe stored color

A bitmap that stores indexes into a color table is called a palette-indexed bitmap (索引). Some bitmaps have no need for a color table shown as Figure 9-2. For example, if a bitmap uses 24 bits per pixel, that bitmap can store the colors themselves rather than indexes into a color table. The following illustration shows a bitmap that stores colors directly (24 bits per pixel) rather than using a color table. The illustration also shows an enlarged view of the corresponding image. In the bitmap, FFFFFF represents white, FF0000 represents red, 00FF00 represents green, and 0000FF represents blue.

0000FF	0000FF	0000FF	0000FF	0000FF	0000FF	0000FF	0000FF
00FF00	FF0000	FFFFFF	FF0000	FFFFFF	FF0000	FFFFFF	00FF00
00FF00	FFFFFF	FF0000	FFFFFF	FF0000	FFFFFF	FF0000	00FF00
00FF00	FF0000	FFFFFF	FF0000	FFFFFF	FF0000	FFFFFF	00FF00
00FF00	FFFFFF	FF0000	FFFFFF	FF0000	FFFFFF	FF0000	00FF00
00FF00	FF0000	FFFFFF	FF0000	FFFFFF	FF0000	FFFFFF	00FF00
00FF00	FFFFFF	FF0000	FFFFFF	FF0000	FFFFFF	FF0000	00FF00
0000FF	0000FF	0000FF	0000FF	0000FF	0000FF	0000FF	0000FF

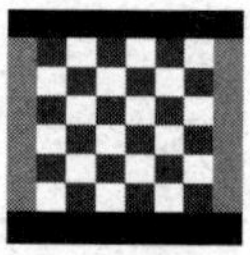

Figure 9-2 A bitmap's directly stored colors

Unit 10　Compilers Principles

Text A　The Science of Code Optimization

The term "**optimization**❶" in **compiler**❷ design refers to the attempts that a compiler makes to produce code that is more **efficient**❸ than the obvious code. "Optimization" is thus a **misnomer**❹, since there is no way that the code produced by a compiler can be **guaranteed**❺ to be as fast or faster than any other code that performs the same task. 注1

In modern times, the optimization of code that a compiler performs has become both more important and more **complex**❻. 注2 It is more complex because processor architectures have become more complex, **yielding**❼ more opportunities to improve the way code **executes**❽. It is more important because massively **parallel**❾ computers require substantial optimization, or their performance suffers by orders of **magnitude**❿. With the likely **prevalence**⓫ of **multi-core**⓬ machines (computers with chips that have large numbers of processors on them), all compilers will have to face the problem of **taking advantage of**⓭ multiprocessor machines.

It is hard, if not impossible, to build a robust compiler out of "hacks." Thus, an extensive and useful theory has been built up around the problem of optimizing code. The use of a rigorous **mathematical**⓮ foundation allows us to show that an optimization is correct and that it produces the desirable effect for all possible inputs. We shall see, starting in Chapter 9, how models such as graphs, **matrices**⓯, and **linear**⓰ programs are necessary if

---

- ❶ 最佳化;最优化
- ❷ 自动编码器;编译器
- ❸ 有效的
- ❹ 误称;用词不当
- ❺ 保证;担保
- ❻ 复杂的;合成物
- ❼ 易受影响的
- ❽ 执行,完成
- ❾ 平行的
- ❿ 大小,量级
- ⓫ 流行,普遍
- ⓬ 多芯的;多核的
- ⓭ 使用,利用
- ⓮ 数学上的,精确的
- ⓯ 模型,矩阵
- ⓰ 线状的

the compiler is to produce well optimized code.

On the other hand, pure theory alone is **insufficient**⑰. Like many real-world problems, there are no perfect answers. In fact, most of the questions that we ask in compiler optimization are un-decidable. One of the most important skills in compiler design is the ability to **formulate**⑱ the right problem to solve. We need a good understanding of the behavior of programs to start with and thorough experimentation and evaluation to **validate**⑲ our intuitions. Compiler optimizations must meet the following design objectives:

The optimization must be correct, that is, preserve the meaning of the compiled program;

The optimization must improve the performance of many programs;

The compilation time must be kept reasonable;

The engineering effort required must be manageable.

It is impossible to **overemphasize**⑳ the importance of correctness. It is trivial to write a compiler that generates fast code if the generated code need not be correct! Optimizing compilers are so difficult to get right that we dare say that no optimizing compiler is completely error-free! Thus, the most important objective in writing a compiler is that it is correct.注3

The second goal is that the compiler must be effective in improving the performance of many input programs. Normally, performance means the speed of the program execution. Especially in embedded applications, we may also wish to **minimize**㉑ the size of the generated code. And in the case of mobile devices, it is also desirable that the code minimizes power consumption. Typically, the same optimizations that speed up execution time also conserve power. Besides performance, usability aspects such as error reporting and debugging are also important.

Third, we need to keep the **compilation**㉒ time short to support a rapid development and debugging cycle. This requirement has become easier to meet as machines get faster. Often, a program is first developed and debugged without program optimizations. Not only is the compilation time reduced, but more importantly, un-optimized programs are easier to debug, because the optimizations introduced by a compiler often obscure the relationship between the source code and the object code.注4 Turning on optimizations in the compiler sometimes exposes new problems **in the source program**㉓; thus testing must again be

---

⑰ 不够的;不足的
⑱ 把……用公式表示
⑲ 证实
⑳ 过分强调
㉑ 使减到最少
㉒ 编译;编辑物
㉓ 在源程序中

performed on the optimized code. The need for additional testing sometimes deters the use of optimizations in applications, especially if their performance is not critical.

Finally, a compiler is a complex system; we must keep the system simple to assure that the engineering and maintenance costs of the compiler are manageable. There is an **infinite**㉔ number of program optimizations that we could implement, and it takes a **nontrivial**㉕ amount of effort to create a correct and effective optimization.注5 We must **prioritize**㉖ the optimizations, implementing only those that lead to the greatest benefits on source programs encountered in practice.

Thus, in studying compilers, we learn not only how to build a compiler, but also the general methodology of solving complex and open-ended problems.注6 The approach used in compiler development involves both theory and experimentation. We normally start by formulating the problem based **on our intuitions**㉗ on what the important issues are.

Words

compilation	n. (名词)	编译
compiler	n. (名词)	自动编码器;编辑人
complex	adj. (形容词)	复杂的;合成物
efficient	adj. (形容词)	有效的
execute	vt. & vi. (动词)	执行,完成
formulate	vt. & vi. (动词)	把……用公式表示;制定
guarantee	n. (名词)	保证;担保
infinite	adj. (形容词)	无限的,无穷的
insufficient	adj. (形容词)	不够的,不足的
linear	adj. (形容词)	线状的
magnitude	n. (名词)	大小,量级
mathematical	adj. (形容词)	数学上的,精确的
matrix	n. (名词)	模型,矩阵
minimize	vt. & vi. (动词)	使减到最少
misnomer	n. (名词)	误称;用词不当
multicore	adj. (形容词)	多芯的;多核的
nontrivial	adj. (形容词)	非平凡的
optimization	n. (名词)	最佳化,最优化

㉔ 无限的,无穷的
㉕ 非平凡的
㉖ 给……优先处理
㉗ 直觉

overemphasize	vt. & vi.（动词）	过分强调
paralle	adj.（形容词）	平行的
prevalence	n.（名词）	流行，普遍
validate	vt.（动词）	证实
yielding	adj.（形容词）	易受影响的

Phrases

in the source program	在源程序
on our intuitions	凭我们的直觉
taking advantage of	利用

Exercises

【Ex1】 **Fill in the blanks according to the text:**

In modern times, the optimization of code that a compiler performs has become both more important and more _____. It is more complex because processor architectures have become more complex, _____ more opportunities to improve the way code _____. It is more important because massively _____ computers require substantial optimization, or their performance suffers by orders of-_____. With the likely prevalence of multi-core machines (computers with chips that have large numbers of processors on them), all compilers will have to face the problem of taking advantage of multiprocessor machines.

【Ex2】 **Translate into Chinese:**

(1) The term "optimization" in compiler design refers to the attempts that a compiler makes to produce code that is more efficient than the obvious code.

(2) In modern times, the optimization of code that a compiler performs has become both more important and more complex.

(3) In fact, most of the questions that we ask in compiler optimization are undecidable.

(4) We need a good understanding of the behavior of programs to start with and thorough experimentation and evaluation to validate our intuitions.

(5) The optimization must be correct, that is, preserve the meaning of the compiled program.

(6) The optimization must improve the performance of many programs.

(7) The compilation time must be kept reasonable.

(8) It is trivial to write a compiler that generates fast code if the generated code need not be correct!

【Ex3】 Choose the best answer：

（1） In modern times, the optimization of code _____ a compiler performs has become both more important and more complex.

    A. that   B. who   C. what   D. where

（2） It is impossible to _____ the importance of correctness.

    A. overemphasize   B. overemphasizing   C. overestimation   D. overestimate

（3） The second goal is that the compiler _____ be effective in improving the performance of many input programs.

    A. should   B. must   C. ought   D. can

（4） We need to keep the compilation time short to support a rapid development and _____ cycle.

    A. debug   B. debugged   C. debugging   D. debugs

（5） In studying compilers, we learn not only how to build a compiler, but also the general methodology of _____ complex and _____ problems.

    A. solving, open-ended   B. solved, open-ended

    C. solving, open-ending   D. solved, open-ending

批 注

注1　Optimization 为主语（优化）；thus a misnomer 为表语；since…引导原因从句；that…修饰 no way；the code produced by a compiler 为从句主语。

注2　That a compiler performs 从句修饰主语 the optimization of code；

Has 为谓语；become…complex 为宾语。

注3　Optimizing compilers 为主语；are 是谓语；so…that 引导状语从语；no optimizing compiler 是从句主语；is 是从句谓语。

注4　主语：compiler；谓语：hides；distributing the computation 分词引导独立结构，与 because 引导的状语从句并列。这句话的含义是：这不只缩短了编译时间，而且更重要的是，因为由编译器引入的优化通常使源代码与目标代码之间的关系变得模糊，所以未经优化的程序更容易调试。

注5　这句话的含义是：我们可以进行无限多的程序优化，而且创造一个正确有效的优化需要很多的努力。

注6　这句话的含义是：因此，研究编译器时，我们不仅要学习如何建立一个编译器，还要学习解决复杂和开放式问题的一般方法。

Text B　Optimizations for Computer Architectures

    The rapid **evolution**❶ of computer architectures has also led to an **insatiable**❷ demand for new compiler technology. Almost all high-performance systems take advantage of the

---

❶ 发展
❷ 无法满足的

same two basic techniques: **parallelism**❸ and memory **hierarchies**❹. Parallelism can be found at several levels: at the instruction level, where **multiple**❺ operations are executed **simultaneously**❻ and at the processor level, where different threads of the same application are run on different processors. Memory hierarchies are a response to the basic **limitation**❼ that we can build very fast storage or very large storage, but not **storage**❽ that is both fast and large.

All modern microprocessors **exploit**❾ instruction-level parallelism. However, this parallelism can be hidden from the programmer. Programs are written as if all instructions were executed in **sequence**❿; the hardware **dynamically**⓫ checks for dependencies in the sequential instruction stream and issues them in parallel when possible. In some cases, the machine includes a hardware **scheduler**⓬ that can change the instruction ordering to increase the parallelism in the program. Whether the hardware reorders the instructions or not, compilers can **rearrange**⓭ the instructions to make instruction-level parallelism more effective.

Instruction-level parallelism can also appear **explicitly**⓮ in the instruction set. VLIW (Very Long Instruction Word) machines have instructions that can issue multiple operations in parallel. The Intel IA64 is a well-known example of such architecture. All high-performance, general-purpose microprocessors also include instructions that can operate on a vector of data at the same time. Compiler techniques have been developed to **generate**⓯ code **automatically**⓰ for such machines from sequential programs.

Multiprocessors have also become prevalent; even personal computers often have multiple processors. Programmers can write **multithreaded**⓱ code for multiprocessors, or parallel code can be automatically generated by a compiler from conventional sequential programs. Such a compiler hides from the programmers the details of finding parallelism in a

---

❸ 平行
❹ 分级
❺ 多样的;复合的
❻ 同时发生地
❼ 限制
❽ 存储
❾ 利用
❿ 顺序,程式
⓫ 动态地
⓬ 调度程序
⓭ 重新排列
⓮ 明确地
⓯ 生成
⓰ 自动地
⓱ 多线程的

program, distributing the computation across the machine, and minimizing **synchronization**❽ and communication among the processors.^{注1} Many scientific-computing and engineering applications are computation-intensive and can benefit greatly from parallel processing. Parallelization techniques have been developed to translate automatically sequential scientific programs into multiprocessor code.

A memory hierarchy **consists of**❾ several levels of storage with different speeds and sizes, with the level closest to the processor being the fastest but smallest. The average memory-access time of a program is reduced if most of its accesses are satisfied by the faster levels of the hierarchy. Both parallelism and the existence of a memory hierarchy improve the potential performance of a machine, but they must be **harnessed**❿ effectively by the compiler to deliver real performance on an application.

Memory hierarchies are found in all machines. A processor usually has a small number of registers consisting of **hundreds of**㉑ bytes, several levels of caches containing **kilobytes**㉒ to megabytes, physical memory containing megabytes to gigabytes, and finally secondary storage that contains **gigabytes**㉓ and beyond. Correspondingly, the speed of accesses between **adjacent**㉔ levels of the **hierarchy**㉕ can differ by two or three orders of magnitude. The performance of a system is often limited not by the speed of the processor but by the performance of the memory subsystem.^{注2} While compilers traditionally focus on optimizing the processor execution, more **emphasis**㉖ is now placed on making the memory hierarchy more effective.

Using registers effectively is probably the single most important problem in optimizing a program. Unlike registers that have to be managed explicitly in software, caches and physical memories are hidden from the instruction set and are managed by hardware. It has been found that cache-management policies implemented by hardware are not effective in some cases, especially in scientific code that has large data structures (arrays, typically). It is possible to improve the effectiveness of the memory hierarchy by changing the layout of the data, or changing the order of instructions accessing the data. We can also change the layout of code to improve the effectiveness of instruction caches^{注3}.

---

❽ 同时性
❾ 由……组成
❿ 治理
㉑ 数以百计
㉒ 千字节
㉓ 千兆字节
㉔ 邻近的
㉕ 层次结构
㉖ 强调

Words

automatically	*adv.*（副词）	自动地
dynamically	*adv.*（副词）	动态地；充满活力地；不断变化地
emphasis	*vt. & vi.*（动词）	强调
evolution	*n.*（名词）	发展
explicitly	*adv.*（副词）	明确地
exploit	*vt. & vi.*（动词）	利用
generate	*vt. & vi.*（动词）	生成，产生
gigabytes	*n.*（名词）	千兆字节
harnessed	*vt.*（动词）	治理
hierarchy	*n.*（名词）	分级
insatiable	*adj.*（形容词）	无法满足的
kilobyte	*n.*（名词）	千字节
limitation	*n.*（名词）	限制
multiple	*adj.*（形容词）	多样的；复合的
multithreaded	*adj.*（形容词）	多线程的
parallelism	*n.*（名词）	平行
rearrange	*vt.*（动词）	重新排列
scheduler	*n.*（名词）	调度程序
sequence	*n.*（名词）	顺序，程式
simultaneously	*adv.*（副词）	同时发生地
storage	*n.*（名词）	存储
synchronization	*n.*（名词）	同时性

Phrases

consists of	由什么组成
hundreds of	数以百计的

批 注

注1　Such a compiler 为句子的主语，finding parallelism in a program, distributing the computation across the machine, and minimizing synchronization and communication among the processors 为并列结构。这句话的含义是：这种编译器对程序员隐藏了在程序中找到并行结构的细节，将计算分布在机器中，并最大限度地减少处理器之间的同步和通信。

注2　这句话的含义是：一个系统的性能往往不是被处理器的速度限制，而是被记忆子系统的性能限制。

注3　我们可以通过改变数据的布局或者改变访问数据的指令的顺序改善内存层次结构的有效性,也可以通过改变代码布局改善指令高速缓存的有效性。

Associated Reading

Introducing "TINY"

In the last installment, I showed you the general idea for the top-down development of a compiler. I gave you the first few steps of the process for compilers for Pascal and C, but I stopped far short of pushing it through to completion. The reason was simple: if we're going to produce a real, functional compiler for any language, I'd rather do it for KISS, the language that I've been defining in this tutorial series.

In this installment, we're going to do just that, for a subset of KISS which I've chosen to call TINY.

The process will be essentially that outlined in Installment IX, except for one notable difference. In that installment, I suggested that you begin with a full BNF description of the language. That's fine for something like Pascal or C, for which the language definition is firm. In the case of TINY, however, we don't yet have a full description, we seem to be defining the language as we go. That's OK. In fact, it's preferable, since we can tailor the language slightly as we go, to keep the parsing easy. So in the development that follows, we'll actually be doing a top-down development of BOTH the language and its compiler. The BNF description will grow along with the compiler.

In this process, there will be a number of decisions to be made, each of which will influence the BNF and therefore the nature of the language. At each decision point I'll try to remember to explain the decision and the rationale behind my choice. That way, if you happen to hold a different opinion and would prefer a different option, you can choose it instead. You now have the background to do that. I guess the important thing to note is that nothing we do here is cast in concrete. When YOU'RE designing YOUR language, you should feel free to do it YOUR way.

Many of you may be asking at this point: Why bother starting over from scratch? We had a working subset of KISS as the outcome of Installment VII (lexical scanning). Why not just extend it as needed? The answer is threefold. First of all, I have been making a number of changes to further simplify the program ... changes like encapsulating the code generation procedures, so that we can convert to a different target machine more easily. Second, I want you to see how the development can indeed be done from the top down as outlined in the last installment. Finally, we both need the practice. Each time I go through this exercise, I get a little better at it, and you will, also.

Many years ago there were languages called Tiny BASIC, Tiny Pascal, and Tiny C, each of which was a subset of its parent full language. Tiny BASIC, for example, had only single-character variable names and global variables. It supported only a single data type. Sound familiar? At this point we have almost all the tools we need to build a compiler like that. Yet a language called Tiny-anything still carries some baggage inherited from its parent language. I've often wondered if this is a good idea. Granted, a language based upon some parent language will have the advantage of familiarity, but there may also be some peculiar syntax carried over from the parent that may tend to add unnecessary complexity to the compiler. (Nowhere is this truer than in Small C.)

I've wondered just how small and simple a compiler could be made and still be useful, if it were designed from the outset to be both easy to use and to parse. Let's find out. This language will just be called "TINY," period. It's a subset of KISS, which I also haven't fully defined, so that at least makes us consistent (!). I suppose you could call it TINY KISS. But that opens up a whole can of worms involving cuter and cuter (and perhaps more risqué) names, so let's just stick with TINY. The main limitations of TINY will be because of the things we haven't yet covered, such as data types. Like its cousins Tiny C and Tiny BASIC, TINY will have only one data type, the 16-bit integer. The first version we develop will also have no procedure calls and will use single-character variable names, although as you will see we can remove these restrictions without much effort.

The language I have in mind will share some of the good features of Pascal, C, and Ada. Taking a lesson from the comparison of the Pascal and C compilers in the previous installment, though, TINY will have a decided Pascal flavor. Wherever feasible, a language structure will be bracketed by keywords or symbols, so that the parser will know where it's going without having to guess.

One other ground rule: As we go, I'd like to keep the compiler producing real, executable code. Even though it may not DO much at the beginning, it will at least do it correctly.

Finally, I'll use a couple of Pascal restrictions that make sense: All data and procedures must be declared before they are used. That makes good sense, even though for now the only data type we'll use is a word. This rule in turn means that the only reasonable place to put the executable code for the main program is at the end of the listing.

The top-level definition will be similar to Pascal:

```
<program> ::= PROGRAM <top-level decl> <main> '.'
```

Already, we've reached a decision point. My first thought was to make the main block optional. It doesn't seem to make sense to write a "program" with no main program, but it does make sense if we're allowing for multiple modules, linked together. As a matter of fact, I intend to allow for this in KISS. But then we begin to open up a can of worms that I'd rather leave closed for now. For example, the term "PROGRAM" really becomes a misnomer.

The MODULE of Modula-2 or the Unit of Turbo Pascal would be more appropriate. Second, what about scope rules? We'd need a convention for dealing with name visibility across modules. Better for now to just keep it simple and ignore the idea altogether.

There's also a decision in choosing to require the main program to be last. I toyed with the idea of making its position optional, as in C. The nature of SK*DOS, the OS I'm compiling for, make this very easy to do. But this doesn't really make much sense in view of the Pascal-like requirement that all data and procedures be declared before they're referenced. Since the main program can only call procedures that have already been declared, the only position that makes sense is at the end, a Pascal.

Given the BNF above, let's write a parser that just recognizes the brackets:

```
{--------------------------------------------------}
{  Parse and Translate a Program }
procedure Prog;
begin
    Match('p');
    Header;
    Prolog;
    Match('.');
    Epilog;
end;
{--------------------------------------------------}
```

The procedure Header just emits the startup code required by the assembler:

```
{--------------------------------------------------}
{ Write Header Info }
procedure Header;
begin
    WriteLn('WARMST', TAB, 'EQU $A01E');
end;
{--------------------------------------------------}
```

The procedures Prolog and Epilog emit the code for identifying the main program, and for returning to the OS:

```
{--------------------------------------------------}
{ Write the Prolog }

procedure Prolog;
begin
    PostLabel('MAIN');
end;

{--------------------------------------------------}
```

```
{ Write the Epilog }

procedure Epilog;
begin
    EmitLn('DC WARMST');
    EmitLn('END MAIN');
end;
{--------------------------------------------------}
```

The main program just calls Prog, and then looks for a clean ending:

```
{--------------------------------------------------}
{ Main Program }

begin
    Init;
    Prog;
    if Look < > CR then Abort('Unexpected data after ''.''');
end.
{--------------------------------------------------}
```

At this point, TINY will accept only one input "program," the null program:

PROGRAM . (or 'p.' in our shorthand.)

Note, though, that the compiler DOES generate correct code for this program. It will run, and do what you'd expect the null program to do, that is, nothing but return gracefully to the OS.

As a matter of interest, one of my favorite compiler benchmarks is to compile, link, and execute the null program in whatever language is involved. You can learn a lot about the implementation by measuring the overhead in time required to compile what should be a trivial case. It's also interesting to measure the amount of code produced. In many compilers, the code can be fairly large, because they always include the whole run-time library whether they need it or not. Early versions of Turbo Pascal produced a 12K object file for this case. VAX C generates 50K!

The smallest null programs I've seen are those produced by Modula-2 compilers, and they run about 200-800 bytes.

In the case of TINY, we HAVE no run-time library as yet, so the object code is indeed tiny: two bytes. That's got to be a record, and it's likely to remain one since it is the minimum size required by the OS.

The next step is to process the code for the main program. I'll use the Pascal BEGIN-block:

    <main> ::= BEGIN <block> END

Here, again, we have made a decision. We could have chosen to require a "PROCEDURE MAIN" sort of declaration, similar to C. I must admit that this is not a bad idea at all ... I don't particularly like the Pascal approach since I tend to have trouble locating the main program in a Pascal listing. But the alternative is a little awkward, too, since you have to deal with the error condition where the user omits the main program or misspells its name. Here I'm taking the easy way out.

Another solution to the "where is the main program" problem might be to require a name for the program, and then bracket the main by

```
BEGIN <name>
END <name>
```

Similar to the convention of Modula 2. This adds a bit of "syntactic sugar" to the language. Things like this are easy to add or change to your liking, if the language is your own design.

To parse this definition of a main block, change procedure Prog to read:

```
{---------------------------------------------}
{  Parse and Translate a Program }
procedure Prog;
begin
    Match('p');
    Header;
    Main;
    Match('.');
end;
{---------------------------------------------}
```

and add the new procedure:

```
{---------------------------------------------}
{ Parse and Translate a Main Program }

procedure Main;
begin
    Match('b');
    Prolog;
    Match('e');
    Epilog;
end;
{---------------------------------------------}
```

Now, the only legal program is:

PROGRAM BEGIN END . (or 'pbe.')

Aren't we making progress??? Well, as usual it gets better. You might try some deliberate errors here, like omitting the 'b' or the 'e', and see what happens. As always, the compiler should flag all illegal inputs.

DECLARATIONS

The obvious next step is to decide what we mean by a declaration. My intent here is to have two kinds of declarations: variables and procedures/functions. At the top level, only global declarations are allowed, just as in C.

For now, there can only be variable declarations, identified by the keyword VAR (abbreviated 'v'):

```
<top-level decls> ::= ( <data declaration> )*
<data declaration> ::= VAR <var-list>
```

Note that since there is only one variable type, there is no need to declare the type. Later on, for full KISS, we can easily add a type description.

The procedure Prog becomes:

```
{--------------------------------------------}
{  Parse and Translate a Program }

procedure Prog;
begin
    Match('p');
    Header;
    TopDecls;
    Main;
    Match('.');
end;
{--------------------------------------------}
Now, add the two new procedures:
{--------------------------------------------}
{ Process a Data Declaration }

procedure Decl;
begin
    Match('v');
    GetChar;
end;
{--------------------------------------------}
{ Parse and Translate Global Declarations }
procedure TopDecls;
begin
    while Look <> 'b' do
        case Look of
```

```
                'v': Decl;
        else Abort('Unrecognized Keyword ''' + Look + '''');
        end;
end;
{---------------------------------------------}
```

Note that at this point, Decl is just a stub. It generates no code, and it doesn't process a list ... every variable must occur in a separate VAR statement.

OK, now we can have any number of data declarations, each starting with a 'v' for VAR, before the BEGIN-block. Try a few cases and see what happens.

Unit 11　Operating System

Text A　Operating System Overview

1. What is an operating system

An operating system is software, consisting of programs and data that runs on computers and manages the computer hardware and provides common services for efficient execution of various application software. 注1

For hardware functions such as input and output and memory allocation, the operating system acts as an **intermediary**❶ between application programs and the computer hardware, although the application code is usually executed directly by the hardware, but will frequently call the OS or be interrupted by it. 注2 Operating systems are found on almost any device that contains a computer—from **cellular**❷ phones and video game consoles to supercomputers and web servers.

2. History of operating system

In the early 1950s, a computer could execute only one program at a time. Each user had sole use of the computer and would arrive at a **scheduled**❸ time with program and data on punched paper cards and tape. The program would be loaded into the machine, and the machine would be **set to**❹ work until the program completed or crashed. Programs could generally be **debugged**❺ via a front **panel**❻ using **toggle**❼ switches and panel lights. It is said that Alan Turing was a master of this on the early Manchester Mark 1 machine, and he was already **deriving**❽ the **primitive**❾ conception of an operating system from the principles of the Universal Turing machine. 注3

Later machines came with libraries of software, which would be linked to a user's program to assist in operations such as input and output and generating computer code from

---

❶ 中介
❷ 蜂窝状的
❸ 预定
❹ 设置,设定
❺ 调试
❻ 面板
❼ 拴牢,系紧
❽ 源于;得自
❾ 原始的;远古的

human-readable symbolic code. This was the **genesis**⑩ of the modern-day operating system. However, machines still ran a single job at a time. At Cambridge University in England the job queue was at one time a washing line from which tapes were hung with different colored **clothes-pegs**⑪ to indicate **job-priority**⑫.

3. Mainframes

Through the 1950s, many major features were pioneered in the field of operating systems, including batch processing, input/output interrupt, buffering, multitasking, spooling, runtime libraries, link-loading, and programs for sorting records in files. These features were included or not included in application software at the option of application programmers, rather than in a separate operating system used by all applications. In 1959 the operating system was released as an integrated utility for the IBM 704, 709, and 7090 **mainframes**⑬.

During the 1960s, IBM's OS/360 introduced the concept of a single OS spanning an entire product line was **crucial**⑭ for the success of System/360. IBM's current mainframe operating systems are distant **descendants**⑮ of this original system and applications written for OS/360 can still be run on modern machines In the mid-'70s, MVS, a descendant of OS/360, offered the first **implementation**⑯ of using RAM as a transparent cache for data.

4. Microcomputer

The first microcomputers did not have the capacity or need for the **elaborate**⑰ operating systems that had been developed for mainframes and minis; minimalistic operating systems were developed, often loaded from ROM and known as monitors. One **notable**⑱ early disk-based operating system was CP/M, which was supported on many early microcomputers and was closely imitated in MS-DOS, which became wildly popular as the operating system chosen for the IBM PC (IBM's version of it was called IBM DOS or PC DOS), its **successors**⑲ making Microsoft. 注4 In the 80's Apple Computer Inc. (now Apple Inc.) abandoned its popular Apple Ⅱ series of microcomputers to introduce the Apple Macintosh

---

⑩ 发生;起源
⑪ 衣服夹子
⑫ 作业优先级
⑬ 大型主机
⑭ 重要的;决定性的
⑮ 后裔;子孙
⑯ 实现,执行
⑰ 精心制作的;详尽的;煞费苦心的
⑱ 值得注意的,显著的;著名的
⑲ 后继

computer with an **innovative**[20] Graphical User Interface (GUI) to the Mac OS.

5. Examples of operating system

(1) Microsoft Windows

Microsoft Windows is a family of **proprietary**[21] operating systems most commonly used on personal computers. It is the most common family of operating systems for the personal computer, with about 90% of the market share currently, the most widely used version of the Windows family is Windows XP, released on October 25, 2001. The newest version is Windows 7 for personal computers and Windows Server 2008 R2 for servers.

(2) UNIX and UNIX-like operating system

Ken Thompson wrote B, mainly based on BCPL, which he used to write UNIX, based on his experience in the multics project. B was replaced by C, and UNIX developed into a large, complex family of inter-related operating systems which have been influential in every modern operating system. The UNIX-like family is a diverse group of operating systems, with several major sub-categories including System V, BSD, and GNU/Linux. <u>The name "UNIX" is a **trademark**[22] of The Open Group which licenses it for use with any operating system that has been shown to **conform to**[23] their definitions.</u> [注5] "UNIX-like" is commonly used to refer to the large set of operating systems which **resemble**[24] the original UNIX.

(3) Mac OS X

Mac OS X is a line of partially proprietary graphical operating systems developed, marketed, and sold by Apple Inc., the latest of which is pre-loaded on all currently shipping Macintosh computers. Mac OS X is the successor to the original Mac OS, which had been Apple's primary operating system since 1984. Unlike its predecessor, Mac OS X is a UNIX operating system built on technology that had been developed at **Next**[25] through the second half of the 1980s and up until Apple purchased the company in early 1997.

<u>The operating system was first released in 1999 as Mac OS X Server 1.0, with a desktop-oriented version (Mac OS X v10.0) following in March 2001.</u>[注6] Since then, six more distinct "client" and "server" editions of Mac OS X have been released, the most recent being Mac OS X v10.6, which was first made available on August 28, 2009. Releases of Mac OS X are named after big cats; the current version of Mac OS X is "Snow Leopard".

---

[20] 革新的,创新的
[21] 所有权;所有人
[22] 商标
[23] 符合
[24] 类似,像
[25] 下一个

(4) Other

Older operating systems which are still used in Niche markets include OS/2 from IBM and Microsoft; Mac OS, the non-UNIX precursor to Apple's Mac OS X; BeOS; XTS-300. Some, most notably Haiku, RISC OS, MorphOS and AmigaOS 4 continue to be developed as **minority**[26] platforms for **enthusiast**[27] communities and specialist applications. Open VMS formerly from DEC, is still under active development by Hewlett-Packard. Yet other operating systems are used almost exclusively in academia, for operating systems education or to do research on operating system concepts. A typical example of a system that fulfills both roles is MINIX, while for example Singularity is used purely for research.

Words

cellular	*adj.* (形容词)	蜂窝状的
crucial	*adj.* (形容词)	重要的;决定性的
debug	*vt. & vi.* (动词)	调试
derive	*vt. & vi.* (动词)	源于;得自
descendant	*n.* (名词)	后裔;子孙
elaborate	*adj.* (形容词)	精心制作的;详尽的
enthusiast	*n.* (名词)	狂热者,热心家
genesis	*n.* (名词)	发生;起源
implementation	*n.* (名词)	实现;履行
innovative	*adj.* (形容词)	革新的,创新的
intermediary	*n.* (名词)	中间人
mainframe	*n.* (名词)	大型主机
minority	*n.* (名词)	少数民族;少数派;未成年
notable	*adj.* (形容词)	值得注意的,显著的;著名的
panel	*n.* (名词)	面板
primitive	*adj.* (形容词)	原始的;远古的
proprietary	*n.* (名词)	所有权;所有人
resemble	*vt. & vi.* (动词)	类似,像
scheduled	*vt. & vi.* (动词)	预定
toggle	*n.* (名词)	拴牢,系紧
trademark	*n.* (名词)	商标

---

[26] 少数
[27] 爱好者

Phrases

clothes-pegs	晒衣用的衣夹
conform to	符合；遵照
job-priority	作业优先级别
set to	开始
shown to	显示
successors	继承人

批 注

注1　句子主干就是 An operating system is software；该句复加在后面的是非限制性定语从句；consisting of programs and data 修饰 software；而 that runs…引导的定语从句是修饰 programs and data。

注2　句子主干是 the operating system acts as an intermediary；Although 引导让步状语从句；It 指代 OS。

注3　It is said 引导的句子是强调句；that 后面则是由 and 连接的两个简单句：Alan Turing was a master…and he was…。

注4　该句主干是 One notable early disk-based operating system was CP/M；which 指代 CP/M. 作引导非限制性定语从句；后面 which became 中的 which 指代 MS-DOS，引导一个非限制性定语从句。

注5　这句话的含义是：UNIX 这个名字是开放自由组织的商标，它用于任何一个操作系统，这个操作系统是符合它的定义的。

注6　这句话的含义是：操作系统的第1版发布时间在1999年，名称是 Mac OS X Server 1.0，在随后的2001年3月发布了一个桌面版的 Mac OS X v10.0。

Exercises

【Ex1】　Answer the questions according to the text：

（1）What is an operating system? How does it work?

（2）What features are pioneered in the field of operating system through the 1950s?

（3）What's the difference between Mainframes and Microcomputers?

（4）What is the Microsoft Windows?

（5）Who wrote the UNIX?

【Ex2】　Translate into Chinese：

（1）In the early 1950s, a computer could execute only one program at a time.

（2）Mac OS X is a line of partially proprietary graphical operating systems developed, marketed, and sold by Apple Inc.

(3) During the 1960s, IBM's OS/360 introduced the concept of a single OS spanning an entire product line was **crucial** for the success of System/360.

(4) Minimalistic operating systems were developed, often loaded from ROM and known as monitors.

(5) The newest version is Windows 7 for personal computers and Windows Server 2008 R2 for servers.

(6) The UNIX-like family is a diverse group of operating systems, with several major sub-categories including System V, BSD, and GNU/Linux.

(7) "UNIX-like" is commonly used to refer to the large set of operating systems which resemble the original UNIX.

(8) A typical example of a system that fulfills both roles is MINIX, while for example Singularity is used purely for research.

【Ex3】 Choose the best answer

(1) _____ is a device that converts images to digital formal.
  A. Copier B. Printer C. Scanner D. Display

(2) _____ are those programs that help find the information you are trying to locate on the WWW.
  A. Windows B. Search Engines C. Web Sites D. Web Pages

(3) An _____ statement can perform a calculation and store the result in a variable so that it can be used later.
  A. executable B. input C. output D. assignment

(4) Each program module is compiled separately and the resulting _____ files are linked together to make an executable application.
  A. assembler B. source C. library D. object

(5) _____ is the conscious effort to make all jobs similar, roytine, and interchangeable.
  A. WWW   B. Informatization
  C. Computerization   D. Standardization

Text B BIOS or CMOS Setup

BIOS is the Acronym for basic input/output system, the built-in software that determines what a computer can do without accessing programs from a disk. On PCs, the BIOS contains all the code required to control the keyboard, display screen, disk drives, serial communications, and a number of miscellaneous functions. All these options will be shown as Figure 11-1、Figure 11-2、Figure 11-3 respectively.

The BIOS is typically placed in a ROM chip that comes with the computer (it is often called a ROM BIOS). This ensures that the BIOS will always be available and will not be damaged by disk failures. It also makes it possible for a computer to boot itself. Because RAM is faster than ROM, though, many computer manufacturers design systems so that the BIOS is copied from ROM to RAM each time the computer is booted. This is known as shadowing.

Because of the wide variety of computer and BIOS manufacturers over the evolution of computers, there are numerous ways to enter the BIOS or CMOS Setup. Below is a listing of the majority of these methods as well as other recommendations for entering the BIOS setup.

Thankfully, computers that have been manufactured in the last few years will allow you to enter the CMOS by pressing one of the below five keys during the boot. Usually it's one of the first three.

F1
F2
DEL
ESC
F10

A user will know when to press this key when they see a message similar to the below example as the computer is booting. Some older computers may also display a flashing block to indicate when to press the F1 or F2 keys.

Press <F2> to enter BIOS setup.

Once you've successfully entered the CMOS setup you should see a screen similar to the Figure 11-1.

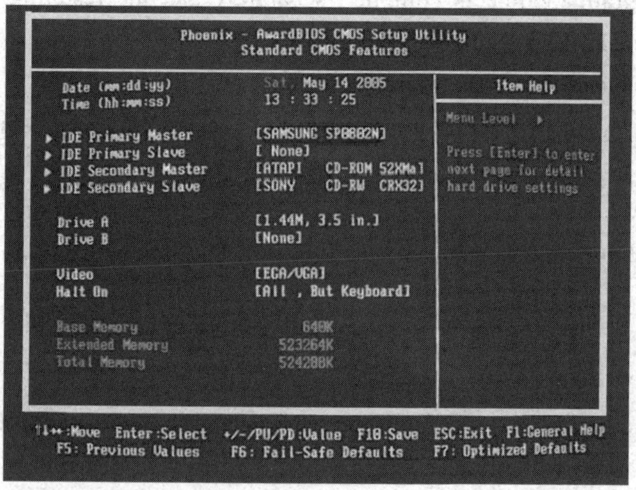

Figure 11-1　BIOS Setup Standard CMOS Features

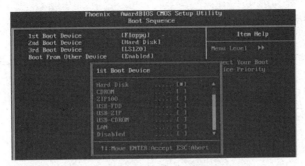

Figure 11-2　BIOS Setup Boot Sequence

Figure 11-3　BIOS Setup Integrated Peripherals

批　注

（1）IDE：集成开发环境

（2）Primary Master：主硬盘

（3）Primary Slave：从硬盘

（4）General Help：一般帮助

（5）Item Help：项目帮助

（6）Boot Sequence：启动顺序

（7）CD-ROM：只读光盘驱动器

（8）USB：通用串行总线

（9）Boot Device：启动设备

（10）LAN：局域网

Words and Phrases

AGP Aperture Size	n.（名词）	AGP 孔径大小
applications	n.（名词）	应用程序
boot device	n.（名词）	启动设备
boot sequence	n.（名词）	启动顺序
CD-ROM	n.（名词）	只读光盘驱动器
delay prior to thermal	n.（名词）	延迟前热
device configuration	n.（名词）	设备设置
DRAM	n.（名词）	动态随机存取存储器

environment subsystems	n. (名词)	环境子系统
general Help	n. (名词)	一般帮助
IDE	n. (名词)	集成开发环境
item help	n. (名词)	项目帮助
kernel	n. (名词)	内核
LAN	n. (名词)	局域网
latency time	n. (名词)	延迟时间
memory hole	n. (名词)	存储器孔
primary master	n. (名词)	主硬盘
primary slave	n. (名词)	从硬盘
service	n. (名词)	服务
shell	n. (名词)	框架
system BIOS cacheable	n. (名词)	系统 BIOS 缓存
system processes	n. (名词)	系统进程
system service dispatcher	n. (名词)	系统服务调度
system thread	n. (名词)	系统线程
USB	n. (名词)	通用串行总线
USB controller	n. (名词)	USB 控制器

Associated Reading

Benefits & Risks of Artificial Intelligence

What is AI?

From SIRI to self-driving cars, artificial intelligence[1] (AI) is progressing rapidly. While science fiction often portrays[2] AI as robots with human-like characteristics, AI can encompass anything from Google's search algorithms to IBM's Watson to autonomous weapons.

Artificial intelligence today is properly known as narrow AI (or weak AI), in that it is designed to perform a narrow task (e.g. only facial recognition or only internet searches or only driving a car). However, the long-term goal of many researchers is to create general AI (AGI or strong AI). While narrow AI may outperform humans at whatever its specific task is, like playing chess or solving equations, AGI would outperform humans at nearly every cognitive task.

Why research AI safety?

In the near term, the goal of keeping AI's impact on society beneficial motivates research in many areas, from economics and law to technical topics such as verification, validity, security and control. Whereas it may be little more than a minor nuisance if your laptop crashes or gets hacked, it becomes all the more important that an AI system does what you want it to do if it controls your car, your airplane, your pacemaker, your automated trading system or your power grid. Another short-term challenge is preventing a devastating arms race in lethal autonomous weapons.

In the long term, an important question is what will happen if the quest for strong AI succeeds and an AI system becomes better than humans at all cognitive tasks. As pointed out by I. J. Good in 1965, designing smarter AI systems is itself a cognitive task. Such a system could potentially undergo recursive self-improvement, triggering an intelligence explosion leaving human intellect far behind. By inventing revolutionary new technologies, such a superintelligence might help us eradicate war, disease, and poverty, and so the creation of strong AI might be the biggest event in human history. Some experts have expressed concern, though, that it might also be the last, unless we learn to align the goals of the AI with ours before it becomes superintelligent.

There are some who question whether strong AI will ever be achieved, and others who insist that the creation of superintelligent AI is guaranteed to be beneficial. At FLI we recognize both of these possibilities, but also recognize the potential for an artificial (the Future of Life Institute) intelligence system to intentionally or unintentionally cause great harm. We believe research today will help us better prepare for and prevent such potentially negative consequences in the future, thus enjoying the benefits of AI while avoiding pitfalls.

How can AI be dangerous?

Most researchers agree that a superintelligent[3] AI is unlikely to exhibit human emotions like love or hate, and that there is no reason to expect AI to become intentionally benevolent or malevolent. Instead, when considering how AI might become a risk, experts think two scenarios most likely:

The AI is programmed to do something devastating: Autonomous weapons are artificial intelligence systems that are programmed to kill. In the hands of the wrong person, these weapons could easily cause mass casualties. Moreover, an AI arms race could inadvertently lead to an AI war that also results in mass casualties. To avoid being thwarted by the enemy, these weapons would be designed to be extremely difficult to simply "turn off," so humans could plausibly lose control of such a situation. This risk is one that's present even with narrow AI, but grows as levels of AI intelligence and autonomy increase.

The AI is programmed to do something beneficial, but it develops a destructive method for achieving its goal: This can happen whenever we fail to fully align the AI's goals with ours, which is strikingly difficult. If you ask an obedient intelligent car to take you to the airport as fast as possible, it might get you there chased by helicopters and covered in vomit, doing not what you wanted but literally what you asked for. If a superintelligent system is tasked with a ambitious geoengineering project, it might wreak havoc with our ecosystem as a side effect, and view human attempts to stop it as a threat to be met.

As these examples illustrate, the concern about advanced AI isn't malevolence but competence. A super-intelligent AI will be extremely good at accomplishing its goals, and if those goals aren't aligned with ours, we have a problem. You're probably not an evil ant-hater who steps on ants out of malice, but if you're in charge of a hydroelectric green energy project and there's an anthill in the region to be flooded, too bad for the ants. A key goal of AI safety research is to never place humanity in the position of those ants.

Why the recent interest in AI safety

Stephen Hawking, Elon Musk, Steve Wozniak, Bill Gates, and many other big names in science and technology have recently expressed concern in the media and via open letters about the risks posed by AI, joined by many leading AI researchers. Why is the subject suddenly in the headlines?

The idea that the quest for strong AI would ultimately succeed was long thought of as science fiction, centuries or more away. However, thanks to recent breakthroughs, many AI milestones, which experts viewed as decades away merely five years ago, have now been reached, making many experts take seriously the possibility of superintelligence in our lifetime. While some experts still guess that human-level AI is centuries away, most AI researches at the 2015 Puerto Rico Conference guessed that it would happen before 2060. Since it may take decades to complete the required safety research, it is prudent to start it now.

Because AI has the potential to become more intelligent than any human, we have no surefire way of predicting how it will behave. We can't use past technological developments as much of a basis because we've never created anything that has the ability to, wittingly or unwittingly, outsmart us. The best example of what we could face may be our own evolution. People now control the planet, not because we're the strongest, fastest or biggest, but because we're the smartest. If we're no longer the smartest, are we assured to remain in control?

FLI's position is that our civilization will flourish as long as we win the race between the growing power of technology and the wisdom with which we manage it. In the case of AI technology, FLI's position is that the best way to win that race is not to impede the former, but to accelerate the latter, by supporting AI safety research.

The top myths about advanced AI

A captivating conversation is taking place about the future of artificial intelligence and what it will/should mean for humanity. There are fascinating controversies where the world's leading experts disagree, such as: AI's future impact on the job market; if/when human-level AI will be developed; whether this will lead to an intelligence explosion; and whether this is something we should welcome or fear. But there are also many examples of of boring pseudo-controversies caused by people misunderstanding and talking past each other. To help ourselves focus on the interesting controversies and open questions—and not on the misunderstandings—let's clear up some of the most common myths shown in Figure 11-4.

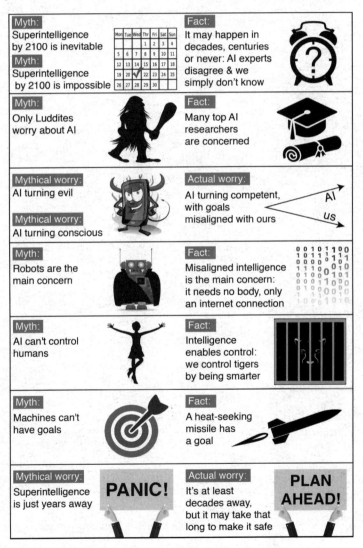

Figure 11-4　Myths and Facts about AI

批 注

(1) artificial intelligence：人工智能
(2) portray：描绘；描述；画像；描画
(3) superintelligent：超智能

附录 A　计算机专业英语主要句型及翻译技巧

专业英语的句型,复合语句、复杂长句较多,由于研究的对象多为客观事物,因此常用形式主语和形式宾语、名词结构和被动语态。另外,在时态上,因为多是介绍事实,故多用现在时。

专业英语句型的学习,在于长期积累,通过句型的学习和总结,有利于掌握英语表达习惯,因此应该加强这方面的积累和训练。在平时的学习总结过程中,建议根据应用场景对句型进行分类,这样便于记忆和应用。本节结合应用的场景对常见专业英语句型进行分类说明。

A.1　计算机专业英语主要句型

1) 基本定义:常见于对新事物进行定义、说明

(1) A connected graph **that** contains no simple circuits **is called** a tree.

包含无简单回路的连接图称为树。

(2) Devices **that** provide input or output to the computer **are called** peripherals.

对计算机提供输入与输出的设备称为外围设备。

(3) I/O **is the means by which** a computer exchanges information with the outside world.

I/O 是计算机通过它可以与外部交换信息的工具。

(4) The term DSL **refers to** digital subscriber line service.

术语 DSL 是指数字用户专线服务。

(5) The base, or radix of a number system **is defined as** the number of different digits.

基数是定义在数字系统中每一位上的不同数字的个数。

(6) We call the INSERT operation on a queue ENQUEUE.

我们称对队列进行的插入操作为入队。

2) 用于比较:常见于为反映事物的某项功能,和具有一定可比性的事物进行比较

(1) **Just as** the CPU controls the computer (in addition to its other functions), the control unit controls the CPU.

同 CPU 控制整个计算机(除了其他功能外)一样,控制单元控制 CPU。

(2) **Compared to** CPU, the speed of the main memory is often slow.

与 CPU 相比,主存的速度通常比较慢。

(3) Buses use less space on a circuit board and require **less** power **than** a large number of direct connections.

与大量的直接连接相比,总线使用较少的电路板空间,耗能更少。

(4) **Unlike** an array, **though**, **in which** the linear order is determined by the array indices, the order in a linked list is determined by a pointer in each object.

不同于数组的顺序由数组下标确定,链表的顺序由每个对象的指针确定。

(5) **Similarly**, both the next field of the tail and the prev field of the head point to nill.

类似地,队尾的 next 域和队头的 prev 域都指向 nill。

3) 功能、用途说明:常用于直接介绍某个对象的作用和功能,或者使用中应注意的事项

(1) A microprocessor typically **performs a sequence of operations to** fetch, decode, and execute an instruction.

典型的微处理器执行取指令、译指令和执行指令等一系列操作。

(2) Computers **have been widely used** in our daily life.

计算机已经广泛用于我们的日常生活。

(3) Its use **is restricted to applications where** high speed is unnecessary.

它仅用于无须高速的情形。

(4) ROM **is typically used to** store the computer's initial start-up instructions.

ROM 常被用于存储计算机的初始启动指令。

(5) Linked lists **provide** a simple, flexible representation **for** dynamic sets.

链表为动态集合提供了简单、灵活的表示形式。

4) 因果关系

(1) **Since** the program counter is (conceptually) just another set of memory cells, **it can be changed by** calculations done in the ALU.

由于程序计数器是另一套存储器单元,因而它能被 ALU 中完成的计算所修改。

(2) An interrupt which can **cause** the computer to stop executing instructions where it was and do something else instead.

中断能使计算机停止正在执行的指令,转而执行其他程序。

(3) If prev[x] = NIL, the element x has no predecessor and **is therefore** the first element, or head, of the list.

prev[x] = NIL 表示元素 x 没有前趋,因此 x 是第一个元素,即队头元素。

(4) An empty list consists of just the sentinel, **since** both next[nill] and prev[nill] can be set to nill.

一个空表只包含标志符,因为指针域 next[nill] 和 prev[nill] 都指向 nill。

5) 组成情况:常用于反映软件或硬件的组成和结构情况

(1) **Inside** each of these parts **are** trillions of small electrical circuits **which can be** turned off or on by means of an electronic switch.

在这些部件的内部是数以万亿计的小的电子电路,这些电子电路能够通过一个电子开关关闭或打开。

（2）**Internally**，The CPU **has** three sections，as shown in Figure 2-3.

CPU 内部有三大分区，如图 2-3 所示。

6）时间描述

常用于表达不同事物、事件发生的时间关系，通常是先、后或同步关系。时间可以是一段时间，也可以是一个时间点。

（1）**During** the fetch **portion of** the instruction **cycle**，the processor first outputs the address of the instruction onto the address bus.

在指令周期的取指阶段，处理器首先将指令的地址输出到地址总线上。

（2）**At the end of** the instruction fetch，the CPU reads the instruction code from the system data bus.

在取指令结束时，CPU 从系统数据总线上读取指令码。

（3）One such mechanism，the instruction pipeline，**allows** the CPU **to** fetch one instruction **while simultaneously** executing another instruction.

这些机制中有一种是指令流水线技术，它允许 CPU 在执行一条指令的同时取出另一条指令。

7）其他句型

（1）Data **is transferred via** the data bus.

数据是通过数据总线传输的。

（2）The INAC instruction of the Relatively Simple CPU **can be executed without** accessing memory or I/O devices。

相对于简单 CPU 的 INAC 指令，不用访问存储器或 I/O 设备即可执行。

（3）We now look at **how** the computer **performs these operations from a system perspective**.

我们从系统的角度看计算机是怎样执行这些操作的。

（4）**This is done entirely within** the microprocessor.

这一步完全在微处理器内完成。

（5）**At this point**，the microprocessor **has fetched** the instruction.

至此，微处理器已经取得了指令。

（6）**When** the READ signal is asserted，the memory subsystem places the instruction code **to be fetched onto** the computer system's data bus.

读信号发出后，存储器子系统就把要取的指令码放到计算机的数据总线上。

（7）It becomes more efficient（**in terms of** minimizing connections）at using buses **rather than** direct connections between every pair of devices.

使用总线比在每个设备对之间直接连接要有效得多（就减少连接数量而言）。

（8）**The desire for** reliability **led** designers **to** use these devices so that they were essentially in one of two states，fully conduction or non-conducting. **A simple analogy may be made between** this type of circuit **and** an electric light.

可靠性的要求使设计者采用这些装置，它们基本上处于两个状态之一：完全导通或截止。在这种电路和电灯之间可以做简单的模拟。

（9）**At any given time** the light（or transistor）is **either** on（conducting）**or** off（not conduction）.

在任意给定的时间里,电灯(或晶体管)处于导通或截止状态之一。

（10）Even after a bulb is old and weak, **it is generally easy** to **tell if it is** on **or** off.

即使电灯泡很旧,一般也很容易区分它是开或关的状态。

（11）**Because of** the large number of electronic parts **used in** computers, it is highly desirable to **utilize them in such a manner that** slight changes in their characteristics **will not affect their performance**.

由于计算机中使用了大量的电子器件,因此非常希望它们拥有这样的特性,即当特性稍有变化时,不至于影响性能。

（12）**Note that** these 32 subsets include T itself and the empty set, which contains no elements at all.

应该注意到,这32个子集中包含T本身和空集,空集是不含任何元素的集合。

（13）**Given** a subset of T, such as S, **we may define** the complement of S **with respect to** a universal set T to consist of precisely those elements of T which are not included in the given subset.

给定T的一个子集,如子集S,可以定义一个关于全集T的S的补集,其中正好包含不在子集S中而在T中的元素。

（14）**Thus**, S **as above defined has** its complement（with respect to T）.

于是,如上定义的集合S就有一个它的补集(相对于集合T)。

（15）**It can be shown that** the finite Boolean algebras are precisely the finite set algebras.

很显然,有限布尔代数一定是有限集合代数。

（16）**While it is possible to use** a different symbol to denote each element of a Boolean algebra, **it is often more useful t**o represent the 2^n elements of a finite Boolean algebra by binary vectors having n components.

虽然可以用不同的符号表示布尔代数中的每一个元素,但最常用的方法是用一个有 n 个分量的二进制向量表示一个有限布尔代数的 2^n 个元素。

（17）**In the remainder of this section**, we **assume that** the lists **with which we are working** are unsorted and doubly linked.

在这一节以后的内容中,假定我们讨论的都是无序的双向列表。

A.2　计算机专业英语翻译方法和技巧

专业英语的文章内容多是描述客观事实,通常结构非常严谨,逻辑性很强,所用专业词汇较多,缩写较多。

1．专业词汇的翻译

（1）选择合适的词义。

有些词语在不同的领域和场合下词义不同，必须结合其所在应用领域翻译。下面用表 A-1 说明这一现象。

表 A-1　部分日常单词在计算机领域中的翻译

单　　词	通　常　情　况	计算机领域
character	性格,字母	字符
memory	记忆力	内存
cell	细胞、监狱	单元格
access	进入	访问,存取
driver	司机	驱动程序
architecture	建筑、结构	体系结构
instruction	说明、教学	指令
cache	藏身处	缓存、高速缓冲存储器
traffic	交通	流量

通常，这些单词的日常翻译和计算机中的翻译存在一定语义上的关联。要能够精准翻译成计算机术语，需要长期积累，结合词义，才能达到精确的专业翻译。

（2）统一翻译习惯。

计算机专业名词的术语，一般结合专业翻译的习惯，统一翻译成同一个词语，见表 A-2。

表 A-2　部分常见易违反习惯的翻译

词　　语	习　惯　翻　译	非习惯的翻译
外围设备	peripheral	external equipment
访问	visit	see
compatible	兼容性	一致性,相容的
response	响应	回答
browser	浏览器	阅读器

在翻译习惯中，特别是对于一些中译英的情况，如果结合字面意思翻译，容易造成词义的偏差较大。在英译翻译中，同样也需要遵循一定的翻译习惯，这就需要加强计算机专业英语的学习，以达到翻译精准、统一的目标。

（3）新词的译法。

当遇到一些专业新词语时，可以根据原词的含义采用适当的方法加以翻译。下面仅以一些曾经出现过的新词汇说明翻译的技巧。

① 音译法

音译法按术语的发音译出，如 bit 翻译为"比特"；又如，Ethernet 翻译为"以太网"。

② 意译法

意译法是结合该词表达的含义进行翻译，如 microcomputer 翻译为"微机"；又如，bare computer 翻译为"裸机"；再如，open source 翻译为"开源"。

2. 语句的翻译技巧

专业英语的翻译，首先需要注重理解语句本身的含义，在明确语句本身含义的前提下，再结合中文和英文的表达习惯和表达方式进行翻译。这主要体现在表达顺序的习惯、词义搭配习惯上。另外，在专业用语上，要体现专业性的表达习惯。翻译是一种艺术，如何精准地找到对应语言的相应词汇，除了掌握一定的翻译方法和技巧外，还需要对语言的表达差异性有深入的理解和长期的翻译经验积累。下面介绍一些专业英语的翻译技巧。

（1）体现实义性，不拘泥于语法。

计算机专业英语在应用领域体现了一种实义性。也就是说，计算机专业英语用词节省、句子精练、采用实义词汇而不拘泥于完整语法，只要能说明问题，阐述中心思想即可。例如：

① 讲究语法但不拘泥于语法。

You will need the user's manual in order to move up from Sunday driver to UNIX speedster.

为了从 UNIX 系统的初学者成为一名熟练者，最好使用用户手册。

② 赋予虚词实际意义。

There is no if in the case 句中的 if 就被赋予了实际意义。句意：这里没有假设的余地。

③ 词性转换翻译

His computer is passworded, I can't use.

他给计算机加了密码，我不能使用。

④ 典故与专用术语

This is the Achilles' heel which makes MS-DOS an unsuitable for integrating information equipment.

直译：这是阿喀留斯的脚跟，它使 MS-DOS 不适合作为信息设备集成的运载体。

参考译文：这是 MS-DOS 的一个致命弱点，使它不适合作为信息设备集成的媒体。

Achilles' heel 是希腊神话中的一个故事，希神阿喀留斯出生后被他的母亲倒提着在冥河水中浸过，除了未浸到水的脚跟外，浑身刀枪不入，所以他的脚跟就成了其致命弱点。

（2）被动语句的翻译。

① 顺序翻译。当原文的被动语句直接翻译成汉语的被动句，又符合汉语习惯时，可顺序翻译。

例如，It can be automatically generated by a compiler.

可翻译为：它可以由(被)一个编译器自动产生。

② 翻译成主动句。当翻译成汉语时，如不符合中文表达习惯，可以译为主动句。这类情况常见于一些词汇，如 expect, require, supply, need 和 cause 等。例如，No return value is required.

可以翻译为：不需要返回数据。

(3) 固定习惯译法。

在专业英语中有一些固定结构，可以采用已有的习惯译法。例如，It has been widely used in… 可以翻译为：它广泛用于……；又如，It lays a solid foundation for… 可以翻译为：为……打下了坚实的基础；再如，With the development of 可以翻译为：随着……的发展。

(4) 复杂句的翻译。

在专业英语中，复杂句、长句较多，对于复杂句的翻译，首先需要把握好主干部分，即主句，以及主句与从句之间存在的逻辑关系，通过恰当的连词将句子拆开翻译，或将一个复杂句拆分成几个简单句。然后将一些附加成分(如介词短语、分词短语、同位语、定语从句、插入语)翻译在合适的位置，在正确理解语句含义的基础上，注意翻译时语句的表达逻辑紧凑，避免歧义、主语不明确、逻辑混乱等情况发生。下面举例说明该翻译技巧。

例句：The World Wide Web, or Web for short, is one of the Internet's most popular services, providing access to over one billion Web pages, which are documents created in a programming language called HTML and which can contain text, graphics, audio, video, and other objects, as well as "hyperlinks" that permit a user to jump easily from one page to another.

翻译：全球信息网，或简称万维网，是因特网上最流行的服务之一，提供对10亿多网页的访问，这些网页是由一种叫作HTML(超文本链接标示语言)的编程语言生成的文件，它可以包含本文、图形、声频、视频和其他对象，以及允许用户容易地跳跃到其他网页的"超链接"。

分析：首先把握语句的主干部分，即全球信息网是因特网上最流行的服务之一。然后分析"providing access to over one billion Web pages"，这是一个"动词+ing"的方式引导的状语从句，相当于并列谓语，可以直接翻译为："提供对10亿多网页的访问"。再分析语句"which are documents created in a programming language called HTML"，这是一个which引导的非限定性定语从句，可以单独作为一条语句翻译，但须明确主语为"这些网页"。

(5) 翻译的语序。

这是指结合中文语言习惯，将一些以 when、not…until、after 等连词引导的从句，在翻译时结合情况将语句的翻译顺序重新排列，在汉语中常常把句子重心放在后面。如：Without the availability of well-specified and functional user-network interface characteristics and the assurance that the network transport function can be achieved using whatever technique best meets the end user's needs, but resulting in no additional interface problems, these goals of ISDN will not be realized.

该句可以翻译为："在没有规范良好的用户接口特性和不能保证使用任何满足终端

用户需要的技术也可以实现网络传输功能的情况下,就算不会引起别的接口问题,ISDN的这些目标也不可能实现。"

又如:Because of the large number of electronic parts used in computers, it is highly desirable to utilize them in such a manner that slight changes in their characteristics will not affect their performance.

该句可以翻译为:"由于计算机中大量采用了电子器件,强烈要求利用它们的一些特性,即当特性稍有变化时,不至于影响性能。"

A.3 通用英语翻译方法

1. 省译法

这是与增译法对应的一种翻译方法,即删去不符合目标语的思维习惯、语言习惯和表达方式的词,以避免译文累赘。增译法的例句反之即可。例如:

(1) You will be staying in this hotel during your visit in Beijing.

你在北京访问期间就住在这家饭店里。(省译物主代词)

(2) I hope you will enjoy your stay here.

希望你在这儿过得愉快。(省译主语)

(3) The Chinese government has always attached great importance to environmental protection.

中国政府历来重视环境保护工作。(省译名词)

(4) The development of IC made it possible for electronic devices to become smaller and smaller.

集成电路的发展使电子器件可以做得越来越小。(省译形式主语 it)

2. 增译法

根据英汉两种语言不同的思维方式、语言习惯和表达方式,在翻译时增添一些词、短句或句子,以便更准确地表达出原文包含的意义。例如:

(1) Indeed, the reverse is true.

实际情况恰好相反。(增译名词)

(2) This is yet another common point between the computers of the two generations.

这是这两代计算机之间的又一个共同点。(增译介词)

(3) Individual mathematicians often have their own way of pronouncing mathematical expressions and in many cases there is no generally accepted "correct" pronunciation.

每个数学家对数学公式常常有各自的读法,许多情况下并不存在一个普遍接受的所谓"正确"读法。(增加隐含意义的词)

(4) It is only when confusion may occur, or where he/she wishes to emphasis the point, that the mathematician will use the longer forms.

只有在可能发生混淆或要强调其观点时,数学家才使用较长的读法。(增加主语)

3. 转换法

在翻译过程中,为了使译文符合目标语的表述方式、方法和习惯,对原句中的词类、句型和语态等进行转换。

(1) 在词性方面,把名词转换为代词、形容词、动词;把动词转换成名词、形容词、副词、介词;把形容词转换成副词和短语。

(2) 在句子成分方面,把主语变成状语、定语、宾语、表语;把谓语变成主语、定语、表语;把定语变成状语、主语;把宾语变成主语。

(3) 在句型方面,把并列句变成复合句,把复合句变成并列句,把状语从句变成定语从句。

(4) 在语态方面,可以把主动语态变为被动语态。

下面通过一些例子说明。

(1) Too much exposure to TV programs will do great harm to the eyesight of children.

孩子们看电视过多会大大损坏视力。(名词转动词)

(2) Thanks to the introduction of our reform and opening policy, our comprehensive national strength has greatly improved.

由于我国实行了改革开放政策,所以我国的综合国力有了明显的增强。(动词转名词)

(3) We don't have much time left. Let's go back.

时间不早了,我们回去吧!(句型转换)

4. 拆句法和合并法

(1) Increased cooperation with China is in the interests of the United States.

同中国加强合作,符合美国的利益。(在主谓连接处拆译)

(2) China is a large country with four-fifths of the population engaged in agriculture, but only one tenth of the land is farmland, the rest being mountains, forests and places for urban and other uses.

中国是一个大国,80%的人口从事农业,但耕地只占土地面积的1/10,其余为山脉、森林、城镇和其他用地。(合译法)

(3) Packet switching is a method of slicing digital messages into parcels called "packets," sending the packets along different communication paths as they become available, and then reassembling the packets once they arrive at their destination.

分组交换是传输数据的一种方法,它先将数据信息分割成许多称为"分组"的数据信息包;当路径可用时,经过不同的通信路径发送;到达目的地后,再将它们组装起来。(将长定语从句拆成几个并列的分句)

5. 正译法和反译法

这两种方法通常用于汉译英,偶尔也用于英译汉。所谓正译,是指把句子按照与汉语相同的语序或表达方式译成英语。所谓反译,则是指把句子按照与汉语相反的语序或表达方式译成英语。正译与反译常常具有同义的效果,但反译往往更符合英语的思维方式和表达习惯,因此比较地道。

(1)你可以从因特网上获得这一信息。

You can obtain this information on the Internet. (正译)

This information is accessible/available on the Internet. (反译)

(2)他突然想到了一个新主意。

Suddenly he had a new idea. (正译)

He suddenly thought out a new idea. (正译)

A new idea suddenly occurred to struck him. (反译)

6. 倒置法

在汉语中,定语修饰语和状语修饰语往往位于被修饰语之前;在英语中,许多修饰语常常位于被修饰语之后,因此翻译时要把原文的语序颠倒过来。倒置法通常用于英译汉,即对英语长句按照汉语的习惯表达法进行前后调换,按意群(指句子中按意思和结构划分出各个成分,每一个成分即称为一个意群)或进行全部倒置,原则是使汉语译句安排符合现代汉语论理叙事的一般逻辑顺序。有时倒置法也用于汉译英。如:

(1) At this moment, through the wonder of telecommunications, more people are seeing and hearing what we say than on any other occasions in the whole history of the world.

此时此刻,通过现代通信手段的奇迹,看到和听到我们讲话的人比整个世界历史上任何其他这样的场合都要多。(部分倒置)

(2)改革开放以来,中国发生了巨大的变化。

Great changes have taken place in China since the introduction of the reform and opening policy. (全部倒置)

7. 包孕法

这种方法多用于英译汉。所谓包孕,是指在把英语长句译成汉语时,把英语后置成分按照汉语的正常语序放在中心词之前,使修饰成分在汉语句中形成前置包孕,但修饰成分不宜过长,否则会形成拖沓或造成汉语句子成分在连接上的纠葛。例如:

(1) IP multicasting is a set of technologies that enables efficient delivery of data to many locations on a network.

IP多信道广播是使数据向网络中许多位置高效传送的一组技术。

(2) What brings us together is that we have common interests which transcend those differences.

使我们走到一起的,是我们有超越这些分歧的共同利益。

8. 插入法

插入法是指把难以处理的句子成分用破折号、括号或前后逗号插入译句中。这种方

法主要用于笔译中,偶尔也用于口译中,即用同位语、插入语或定语从句处理一些解释性成分。例如:

If the announcement of the recovery of Hong Kong would bring about, as Madam put it, "disastrous effects," we will face that disaster squarely and make a new policy decision.

如果说宣布收回香港就会像夫人说的"带来灾难性的影响",那我们将勇敢地面对这个灾难,做出决策。

9. 重组法

进行英译汉时,为了使译文流畅和更符合汉语叙事论理的习惯,在捋清英语长句的结构、弄懂英语原意的基础上,彻底摆脱原文语序和句子形式,对句子进行重新组合。例如:

Decision must be made very rapidly; physical endurance is tested as much as perception, because an enormous amount of time must be spent making certain that the key figures act on the basis of the same information and purpose.

必须把大量时间花在确保关键人物均根据同一情报和目的行事,而这一切对身体的耐力和思维能力都是一大考验。因此,一旦考虑成熟,决策者就应迅速做出决策。

10. 综合法

综合法是指单用某种翻译技巧无法译出时,着眼篇章,以逻辑分析为基础,同时使用转换法、倒置法、增译法、省译法、拆句法等多种翻译技巧的方法。例如:

Behind this formal definition are three extremely important concepts that are the basis for understanding the Internet: packet switching, the TCP/IP communications protocol, and client/server computing.

在这个正式的定义背后,隐含着3个极其重要的概念:分组交换、TCP/IP(传输控制协议/网际协议)通信协议和客户端/服务器计算技术,它们是理解因特网的基础。

11. 长句翻译应用

正确分析句子的结构是正确翻译的基础,为了做好专业英语长句的翻译,要先从长句的句法结构分析入手。专业英语的结构复杂的长句,其句法结构主要包括4个方面:①大量使用名词化结构以及垂悬结构压缩句子长度;②常用先行词 it 结构保持文风凝重平稳;③常用平行结构使层次清晰;④使用圆周句结构,从句交错重叠。

英语中的长句通常结构复杂,修饰成分多而不定,看上去无章可循。但是,依照西方人重逻辑和条理,多线性思维的特点,英语长句多采取分层结构,讲究主次,所以翻译时应根据英汉表达方式的不同,采取不同的翻译方法将英语长句自然地用汉语表达出来。通常采用的方法有顺译法、逆译法、分译法以及综合译法等。下面结合一些例子对这些方法进行分析。

(1) Moving around the nuclear are extremely tiny particles, called electrons, which revolve around the nuclear in much the same way as the nine planets do around the sun. (参考译文:围绕原子核运动的是一些极其微小的粒子,称为电子,这些电子围绕原子核旋

转,正像九大行星围绕太阳旋转一样。)

本句中的"Moving around the nuclear"为动名词短语,在句中作主语;"Which revolve…sun"为非限制性定语从句,由关系代词which引导,修饰先行词electrons。经过分析,此句结构清晰,各部分基本上按逻辑关系排列。这与汉语的表达习惯基本一致,可以按照英语原文的顺序翻译成汉语,对这类长句的翻译采用的是顺译法,即先把句子拆分成意思单位,然后依次译出,需要变化的就随之变化,使之更符合汉语的表达习惯。

(2) A material which has the property of elasticity will return to its original size and shape when the force producing strain are removed. (参考译文:如果把产生应变的力去掉,具有弹性的材料就会恢复到它原来的体积和形状。)

本例中由when引导的时间状语从句,按照汉语的表达习惯位于主句之后的状语从句、主语从句等,翻译时往往要放在主句之前,这种译法叫作逆译法,即不限于原文的形式,逆着原文的顺序翻译,把英语长句中的首句置于汉语句的末句。

(3) The loads a structure is subjected are divided into dead loads, which include the weights of all the parts of the structure, and live loads, which are due to the weights of people, movable equipment, etc. (参考译文:一个结构物受到的荷载可以分为静载荷和动载荷两种。静载荷包括该结构物各部分的重量;动载荷是由人、可移动的设备等的重量引起的。)

本句有两个并列的宾语,分别是由which引导的非限制性定语从句。从此例可以看出,英语长句中,主语和从句或修饰成分的关系并不紧密,而汉语就有常常使用短句表达的习惯。把长句中的从句和修饰成分译成短句分开翻译,这种方法叫作分译法。

(4) More particularly, this invention relates to a method and apparatus for forming a film of metal oxides continuously on the surface of ribbon glass by spring thereon a solution of metal compounds at a point in the neighborhood of the inlet to a lehr or the inside thereof when the ribbons glass is being conveyed to the lehr after it has been formed from molten glass. (更具体地说,本发明涉及在带玻璃表面上连续地喷涂一层金属氧化膜的方法和装置。当玻璃由液态成型为带玻璃之后,在其被送往退火窑时,从退火窑入口附近或入口里将金属化合物喷到玻璃表面。)

翻译英语长句时,并不是单纯地使用一种翻译方法,而是要求我们综合使用各种方法,以便把英语原文翻译成通顺忠实的汉语句。在该例,介词by的前后句子表达的是两层意思。前面的语序和汉语相同,可采用顺译法;后面的语序与汉语完全相反,可采用逆译法。这样,根据句子及修饰语的主次、逻辑、时间先后,有顺有逆地综合处理的翻译方法叫作综合译法。

附录 B 计算机专业英语的特点

B.1 计算机专业英语的词汇特点

专业词汇语法的特点主要表现在专业术语较多,合成义单一。计算机行业中不断涌现出新构造出来的词,如 unformat、uninstall 和 blog 等,这些词往往在词典中查不到。但这些单词的形成往往体现出一定的规律,这就是构词法。掌握了构词法,对于计算机英语的学习,就可以达到举一反三、见词识义的效果。常用的构词法有合成法、转化法及词缀法等。

1. 合成法

合成法是指把两个或两个以上独立的词按照一定次序排列构成新词的方法。由这种方法构成的词具有灵活、机动、善变、言简意赅的特点,例如 network(网络)、backup(备份,已另外储存以备份计算机发生故障时间的备份程序或文件)、database(数据库,指互相关联的事实或数字并已计算机化的表列)、dialogue box(对话框,供用户与计算机进行一系列问答操作的栏面)、harddisk(硬盘)、hardware(硬件)、software(软件)、download(下载)、laptop(笔记本式计算机、便携式机)、offline(脱机)、homepage(主页)、backbone(主干网)、login(登录,输入用户名和密码,以使计算机验证你是计算机或网络的合法用户,并允许你使用资源)和 logoff(注销,停止使用计算机或网络时所执行的步骤,通常是一条单独的命令)。

常见复合词有:

- -based:基于,以……为基础。例如,rate-based:基于速率的,credit-based:基于信誉的,file-based:基于文件的。
- -oriented:面向……的。例如,object-oriented:面向对象的,market-oriented:市场导向,process-oriented:面向进程的。
- info-:信息,与信息有关的。例如,info-channel:信息通道,info-tree:信息树,info-world:信息世界。
- event-:事件的。例如,event-driven:事件驱动的,event-oriented:面向事件的,event-based:基于事件的。
- -free:自由的,无关的。例如,paper-free:无纸的,jumper-free:无跳线的,lead-free:无线的。
- -centric:以……为中心的。例如,client-centric:以客户为中心的,user-centric:以用户为中心的,host-centered:以主机为中心的。

2. 转化法

转化法是指一种词类转换为另一种词类,转换后的词义往往与原来词义有密切联系。例如,reject(n. 废品)转化为 to reject(v. 拒不接受),Format(n. 格式)转化为 to format(v. 格式化)。

在计算机英语材料里,最令人头疼的是我们会碰到一些在一般英语里已经认识而其实际含义却与已知意义大相径庭的短语,此时稍不留神,便会误入歧途,落进"陷阱",出现差之毫厘、失之千里的误解。

例如:character(特征、性格、角色)——字符,menu(菜谱)——菜单、选单,program(节日)——程序、编程,operation(手术)——操作,display(显示)——显示器,storage(贮存)——存储器,track(跑道)——磁道,save(挽救、节省)——存盘,mouse(老鼠)——鼠标,memory(记忆)——内存,bus(公共汽车)——总线,driver(驾驶员)——驱动程序,monitor(班长)——显示器,enter(进入)——回车,key(钥匙)——按键等。

在计算机英语中,这样的词汇很多,都是我们熟悉的初、中级词汇,不过含义已经有所不同。所以,在某种意义上,学习计算机英语,词汇应该不是"记忆"的问题,而是"转义"的问题。大部分单词是熟悉的,只在学习过程中将它对应到计算机专业课程中已经熟知的专有词义就可以了。

3. 词缀法

词缀法是指在一个单词的前面或后面加上一定的词缀构成一个新词的方法。按照词缀在词中的位置,可分为前缀法和后缀法。在计算机英语中,构词能力强的前缀有

- re:reboot——重新起动,retry——重试,Recycle bin——回收站,refresh——刷新,rename——重命名,replay——回放。
- pre:preview——预览,preset——预置,preprocessor——预处理。
- cyber:cyberspace——计算机网络系统,cybershop——网络商店,cyberchat——网络聊天。
- inter:Internet——因特网,interactive——交互式的,interface——接口,界面。
- micro:Microsoft——微软公司,microcomputer——微机,microprocessor——微处理器。
- multi:multimedia——多媒体,multitasking——多任务。

有关的后缀词:

- able:erasable——可擦除,programmable——可编程。
- or:accelerator——快捷键,calculator——计算器。
- ise/ize:digitize——数字化,maximize——最大化,normalize——标准化。
- er:driver——驱动程序,browser——浏览器,server——服务器,scanner——扫描仪,hacker——黑客。

4. 缩写词

缩写词是指取词组中每个词的第一个字母组合构成的新词(可参考附录 C)。例如,

ALU(Arithmetic Logic Unit,算术逻辑运算单元)、LAN(Local Area Network,局域网)等。

5. 借用词

借用词是指借用公共英语及日常生活用语中的词汇表达专业含义。借用词一般来自厂商名、商标名、产品代号名、发明者名、地名等,也可将普通公共英语词汇演变成专业词义实现,也有对原来词汇赋予新的意义的。

例如:cache——高速缓存、隐藏处所、隐藏的粮食或物资、贮藏物,Firewall——防火墙、隔火墙,Flag——标志、状态、旗、标记,mail-bomb/junk-mail——邮件炸弹,bomb——炸弹。

B.2　计算机专业英语的语法特点

专业英语的语法特点:常用形式主语和形式宾语,名词结构和被动语态使用频繁,复杂长句多,复合句多,每句包含信息量大。另外,在时态上,叙述事实多用现在时。

(1) 频繁使用陈述句,时态多用现在时。由于专业英语多描述一些客观事实,因此常用陈述句表达方式描述。如 The text you enter into your document is usually just the raw material for your finished product.

(2) 多用被动语态。由于通常表达的对象是物,因此所用语态为被动语态。如 The main memory of most computer is composed of RAM.

(3) 多使用祈使句。常用于命令、劝告、建议等介绍说明、阐述实例等,其主语往往省略。如 Drag the text to the space in front of the roof.

(4) 定语后置的使用较多。如 Java is a language very difficult to learn.

(5) 动词、动词的非谓语形式出现频率比较高。论述的重点往往是事实现象或过程,不涉及有关的人,故动词的被动语态形式也有很高的出现频率。如:

MENU—has not been defined(MENU—还未定义);Record appears to be out of place(记录位置不对);Memory length is not supported and was ignored(不支持存储器长度并忽略它);The network is inactive or you are not connected properly(此网络是禁用的或未正常连接);The new password has been used previously, please try again(新口令先前已用过,请重试);The partition was cleared and the initialization procedure shold be restarted(清除该盘分区,再重新启动初始化过程)。动词的非谓语形式:Extended Dos partition created(扩展 DoS 分区已经建立),Expected network number missing(要求的网络号遗漏);Error loading operatiin system(装入操作系统时出错);No partition to delete(没有可删除的分区);Insufficient memory to complete fill(内存不够,不能完成填充工作)。

B.3　计算机专业英语的修辞特点

计算机专业英语属于科技英语,具有科技英语的修辞特点。文体一般是一种客观的叙述,用于描述事实、记录试验、阐明规律或探讨理论。不论是科学论文、科普论文,还是技术文本,都需要把科学道理说清楚,具有平铺直叙、简洁、确切的特点。科技人员注重的是科学事实和逻辑概念等,因而其文体十分严谨,修辞比较单调。许多语言学者认为,科技语言应当避免抒情、幽默、比喻以及其他任何带有主观色彩的语言,以免使读者产生行文浮华、内容失真之感。但修辞格未必就会削弱语言内容的客观性,相反,恰当的修辞格能起到润滑剂的作用,减少晦涩,使行文流畅生动、形象易懂、富有美感。正因如此,在科技文献中采用适当的修辞格(如明喻、暗喻、提喻、拟人等),可以提升读者对文章内容的理解。在计算机专业英语阅读和写作时,要正确体会这一点。

1. 词语修辞

词语修辞常用的具体修辞格有隐喻(metaphor)、拟人(personification)、夸张(hyperbole)、仿拟(parody)、矛盾(oxymoron)、移就(transferred epithet)、对照(contrast)、借代(metonymy)及提喻(synecdoche)等,这些修辞格在科技英语的翻译中均有体现,在计算机专业英语中也不少见。下面以拟人为例介绍词语修辞。

拟人赋予各种事物以人的特征、思想或活动,或带上人的属性,形象解释科技知识,使内涵化虚为实,格调轻松,明朗自然。译文应明确爽洁,直截了当,运用于其中也产生异曲同工之妙,增加审美想象力。例如:

(1) While many Californians dream of leaving their home state of California, clams from Asia, a slug from New Zealand and a jellyfish from the Black Sea have decided to buck the trend and take up residence in San Francisco Bay.

译文:正当许多加利福尼亚人梦想着离开他们的家园时,亚洲的一些蛤类、新西兰的一种海参以及生长在黑海的一种水母却拿定主义反潮流,来到旧金山定居(将蛤类等海生物拟人化,"拿定主义反潮流"这几个字把它们的行为描述得更加人性化,用在此处来与句前人的行为进行对比)。

(2) In the chilly month of November, most U.S. production areas don't have harvest fruit. But a team of University of Florida scientists is trying to help growers in Northern Florida produce a crop—by tricking strawberry plants into thinking it's spring time.

译文:寒风凛冽的十一月对美国的大多数水果产地来说,不是收获季节,佛罗里达大学的一个研究小组却正在尝试让佛罗里达北部果农在这时候也有收成——方法是对草莓进行欺骗,让它们觉得春天已经来临了(赋予草莓以生命,"对草莓进行欺骗,让它们觉得春天已经来临了"这几个字让译文读者更能够清楚地明白这个研究小组研究的实际效果)。

(3) The technology of millimeter wave guidance is still in its infancy now.

译文:毫米波制导技术目前才刚刚起步。

(4) Since the early 50s, when naval missiles came of age, the way in which the different navies evaluated the level of importance to be attributed to naval gun has been marked by ups and downs.

译文:自从20世纪50年代初舰载导弹开始服役以来,各国海军对舰炮重要性的评价就各不相同。

例(3)中的短语 in its infancy 和例(4)中的 came of age 一般用来描写人,分别表示"婴儿期"和"成熟,到法定年龄",但这里用来说明"制导技术"和"舰载导弹"的发展情况。

(5) "Because this protein comes from a related fungus, the tree thinks it's under attack," says Hubbes:"It therefore triggers a cascade of defense reactions."

译文:"因为这种蛋白质来自一种相关真菌,所以这棵树认为自己受到攻击",哈布斯说道。"它因此采取一连串的防御性行动。"(作者把树比作人,赋予它思想。这是有生命的事物间的比拟)

(6) NEC says it has designed the robot to be "cute", It can doze off and talks in its sleep, and will start dancing and playing music if it is petted.

译文:日本电气公司称已设计出一种灵巧的机器人,它会打瞌睡和说梦话。要是你拍它一下,它便开始跳舞、演奏音乐(无生命的事物也可拟人化)。

2. 音韵修辞

音韵修辞常用的具体修辞格有押韵和拟声(onomatopoeia)。它们的合理运用能使翻译明快而生动,富有节奏感,充分体现语言内在的音乐美。下面以押韵为例分析音韵修辞。

押韵在英美报刊或科技报告中俯拾皆是。头韵或尾韵都通过重复相同读音或字母,使结构整齐漂亮,音律铿锵有力,增加了和谐与美感。这一重要修辞手段给人以听觉冲击或视觉美感。英语辞格押头韵的手法历来被认为是很难译的,译文要求最好能再现原文鲜明的修辞特点,获得形和声两方面的美。如:

(1) The electronic devices are used in computers as switches that simple turn on and off.

译文一:这些电子器件在计算机中起开关作用,只是打开和关掉而已。

译文二:这些电子器件在计算机中起开关作用,只是开开关关而已。

(2) And soon over the whole surface of the marsh, a great cloud of birds hung screaming and circling in the air.

译文一:很快就有一块乌云似的一大群野鸟在沼泽上空惊叫着,盘旋着。

译文二:很快就有黑压压的一大群野鸟在沼泽地上空惊叫着,盘旋着。

英语的 alliteration(头韵)是英语的骄傲,是英语获得音韵美的台柱,而汉语的叠音词则是另一道更加绚丽的音韵风景。译文一与译文二相比,不难发现译文二因使用了汉语表达的叠音词而朗朗上口,其音韵美又为意境的营造推波助澜,而译文一并非误译,但读来逊色。遇到科技汉英翻译时,译者应该在准确翻译的基础上积极运用头韵,使得译文在形式上更加紧凑。

(3) 搜集资料加以论证,提出理论加以检验,然后归纳整理研究成果——这就是科学工作的全部内容。

译文:Gathering facts, conforming them, suggesting theories, testing them, and organizing findings — this is all the work of science.(词尾-s,-em 交替押韵)

（4）脱发可能由药物、疾病和食物引起。

译文：Hair loss can be triggered by drug, disease and diet.（词头 d-形成视觉押韵效果，读起来也朗朗上口）

3. 结构修辞

结构修辞常用的具体修辞格有 repetition（反复）、catchword repetition（联珠）、chiasmus（回文）、parallelism（排比）、antithesis（反对）、rhetoric question（设问）和 anticlimax（突降）等。下面以排比为例介绍结构修辞。

排比是用两个或两个以上结构相同或形似的句子或句子成分表示相近或相关的内容，反复强调某个观点，使人留下深刻印象。排比读起来音韵悠扬，层层递进，具有流畅的美。译文语言应力求概念准确，活泼生动，行文严谨流畅。如：

（1）The Internet is also fun. You can write to old friends. Or check out their web pages. Enjoy web based soap operas. Laugh at online parodies and jokes. Join in a live Net Event. Chat with other Internet surfers. Pick a fake stock portfolio…

译文：因特网也很好玩。你可以给老朋友写信，或查看他们的网页；欣赏网上的电视连续剧；网上还有模仿的滑稽剧和笑话让你开心；你可以参与实况网络动态节目；与别的因特网冲浪者聊天；模拟跟踪股票价……（原文的排比由一连串短句组成，译文条理清晰、语势如瀑）。

（2）It is the same old story of not being grateful for what we have until we lose it, of not being conscious of health until we are ill.

译文：还是那句老话，失去方知难得，病后倍思健康（原文排比包含两个 of not being…结构，译文娓娓叙来，层层递进）。

（3）Manufacturing processes may be classified as unit production with small quantities being made and mass production with large numbers of identical parts being produced.

译文：制造方法可分为单件生产和批量生产。单件生产是指小批量的生产。批量生产是指生产大量相同的部件（由于汉语重意合，汉语句子间的关系主要由意思本身连接，很少借助连词或介词；而英语重形合，句子之间的结构比较严密，介词和连词（特别是连词）使用很多，句子的各个组成成分和句子与句子之间的关系明确，主次清楚。因此，汉译时不能全盘照搬原文的句子结构，而应按照汉语多用短句的习惯，采用先总后分，讲究对称、排比的原则，达到整齐美的效果）。

（4）Associated with limit loads is the proof factor, which is selected to ensure that if a limit load is applied to a structure the result will not be detrimental to the functioning of the aircraft, and the ultimate factor, which is intended to provide for the possibility of variations in structural strength.

译文：与极限荷载有关的是保险系数和终级系数。选合适的保险系数保证即使极限荷载加在飞机上，飞行也不会受到损害；终极系数则是用来预防结构强度可能发生的变化（译文采用排比结构，用了由非限制性定语从句引导的两个相同结构，显得对称整齐、和谐优美，给人一气呵成之感）。

B.4 计算机专业科技论文的结构特点

1. 科技论文的结构

一篇完整规范的学术论文结构包括 Title、Abstract、Introduction、Method、Result、Discussion、Conclusion 和 Reference 八项内容,这是必不可少的,其他内容则根据具体需要而定,如 Keywords(关键词)、Table of contents(目录)、Nomenclature(术语表)、Acknowledgement(致谢)、Appendix(附录)、Notes(注释)。

2. 科技论文标题的写法

学术文章的标题主要有三种结构:名词性词组(包括动名词)、介词词组、名词/名词词组+介词词组。在人文社会科学领域,有时也用一个疑问句作标题,但一般不用陈述句或动词词组作标题。

1)名词性词组

名词性词组由名词及其修饰语构成。名词的修饰语可以是形容词、介词短语,有时也可以是另一个名词。名词修饰名词时,往往可以缩短标题的长度。以下各标题分别由两个名词词组构成。例如:

latent demand and the browsing shopper(名词词组+名词词组)

cost and productivity(名词+名词)

2)介词词组

介词词组由介词+名词或名词词组构成。如果整个标题就是一个介词词组,一般这个介词是 on,意思是"对……的研究"。例如:

from knowledge engineering to knowledge management(介词词组+介词词组)

on the correlation between working memory capacity and performance on intelligence tests

3)名词/名词词组+介词词组

这是标题中用得最多的结构。例如:

simulation of controlled financial statements(名词+介词词组)

the impact of internal marketing activities on external marketing outcomes(名词+介词词组+介词词组)

diversity in the future work force(名词+介词词组)

models of sustaining human and natural development(名词+介词词组)

标题中的介词词组一般用来修饰名词或名词词组,从而限定某研究课题的范围。这种结构与中文的"的"字结构相似,区别是中文标题中是修饰语在前,中心词在后。英文正好相反,名词在前,而作为修饰语的介词短语在后。例如:

progress on fuel cell and its materials(燃料电池及其材料进展)

4）其他形式

对于值得争议的问题，偶尔可用疑问句作为论文的标题，以点明整个论文讨论的焦点。例如：

Is B2B e-commerce ready for prime time?

Can ERP Meet Your eBusiness Needs?

有的标题由两部分组成，用冒号（:）隔开。一般来说，冒号前面一部分是研究的对象、内容或课题，比较笼统，冒号后面具体说明研究重点或研究方法。这种结构可再分为三种模式。

模式1 研究课题:具体内容。例如：

Microelectronic Assembly and Packaging Technology: Barriers and Needs

The Computer Dictionary Project: an update

模式2 研究课题:方法/性质。例如：

B2B E-Commerce: A Quick Introduction

The Use of Technology in Higher Education Programs: a National Survey

模式3 研究课题:问题焦点。例如：

Caring about connections: gender and computing

3. 英文摘要的写作技巧

英文摘要（Abstract）的写作应用很广。论文摘要是全文的精华，是对一项科学研究工作的总结，是对研究目的、方法和研究结果的概括。

1）摘要的种类与特点

摘要主要有以下四种。

第一种是随同论文一起在学术刊物上发表的摘要。这种摘要置于主体部分前，目的是让读者首先了解一下论文的内容，以便决定是否阅读全文。一般来说，这种摘要在全文完成之后写。字数限制为100~150字。内容包括研究目的、研究方法、研究结果和主要结论。

第二种是学术会议论文摘要。会议论文摘要往往在会议召开之前几个月撰写，目的是交给会议论文评审委员会评阅，从而决定是否能够录用。所以，这种摘要比第一种略为详细，字数为200~300字。在会议论文摘要的开头有必要简单介绍一下研究课题的意义、目的、宗旨等。如果在写摘要时研究工作尚未完成，全部研究结果还未得到，就应在方法、目的、宗旨、假设等方面多花笔墨。

第三种为学位论文摘要。学士、硕士和博士论文摘要一般都要求用中、英文两种语言写。学位论文摘要一般在400字左右，根据需要可以分为几个段落。内容一般包括研究背景、意义、主旨和目的;基本理论依据、基本假设;研究方法;研究结果;主要创新点;简短讨论。不同级别的学位论文摘要，要突出不同程度的创新之处，指出有何新的观点、见解或解决问题的新方法。

第四种是脱离原文而独立发表的摘要。这种摘要更应该具有独立性、自含性、完整性。读者无须阅读全文，便可以了解全文的主要内容。

2）摘要的内容与结构

摘要内容一般包括：

目的（objectives,purposes）：包括研究背景、范围、内容、要解决的问题及解决这一问题的重要性和意义。

方法（methods and materials）：包括材料、手段和过程。

结果与简短讨论（results and discussions）：包括数据与分析。

结论（conclusions）：主要结论，研究的价值和意义等。

概括地说，摘要必须回答"研究什么""怎么研究""得到了什么结果""结果说明了什么"等问题。无论哪种摘要，语言特点和文体风格都相同。首先必须符合格式规范。语言必须规范通顺、准确得体，用词要确切、恰如其分，而且要避免非通用的符号、缩略语、生偏词。另外，摘要的语气要客观，不要做出言过其实的结论。

3）摘要的英文写作风格

英文摘要写作要求句子完整、清晰、简洁；力求用简单句。为避免单调，可改变句子的长度和句子的结构。用过去时态描述作者的工作，因为是过去所做的，但是用现在时态描述所做的结论。

摘要中避免使用动词的名词形式。如：

正："Thickness of plastic sheet was measured"

误："measurement of thickness of plastic sheet was made"

正确使用冠词，既应避免多加冠词，也应避免蹩脚地省略冠词。如：

正："Pressure is a function of the temperature"

误："The pressure is a function of the temperature"；

使用长的、连串的形容词、名词或形容词加名词修饰名词。为打破这种状态，可使用介词短语，或用连字符连接名词词组中的名词，形成修饰单元。例如：

应写为"The chlorine-containing propylene-based polymer of high melt index"，

而不写为"The chlorine containing high melt index-propylene based polymer"

使用短的、简单的、具体的、熟悉的词。不使用华丽的辞藻。使用主动语态，而不使用被动语态。"A exceeds B"读起来要好于"B is exceeded by A"。使用主动语态还有助于避免过多地使用类似于 is、was、are 和 were 这样的弱动词。

构成句子时，动词应靠近主语。避免形如以下的句子：

"The decolorization in solutions of the pigment in dioxane, which were exposed to 10 hr of UV irradiation, was no longer irreversible."

改进的句子应当是：

"When the pigment was dissolved in dioxane, decolorization was irreversible, after 10 hr of UV irradiation."

避免使用既不说明问题，又没有任何含义的短语。例如："specially designed or formulated""The author discusses""The author studied"应删去。

4. 正文

学术论文的正文一般包括 Method、Result、Discussion 三部分。这三部分主要描述研究课题的具体内容、方法,研究过程中所使用的设备、仪器、条件,并如实公布有关数据和研究结果等。Conclusion 是对全文内容或有关研究课题进行的总体性讨论。它具有严密的科学性和客观性,反映一个研究课题的价值,同时提出以后的研究方向。

为了帮助说明论据、事实,正文中经常使用各种图表。最常用的是附图(Figure)和表(Table),此外还有图解或简图(Diagram)、曲线图或流程图(Graph)、视图(View)、剖面图(Profile)、图案(Pattern)等。在文中提到时,通常的表达法为

如图 4 所示　　As (is) shown in Fig. 4,

如表 1 所示　　As (is) shown in Tab. 1,

5. 结论

正文最后应有结论(Conclusions)或建议(Suggestions)。

(1) 关于结论,可用如下表达方式:

① The following conclusions can be drawn from …(由……可得出如下结论)

② It can be concluded that …(可以得出结论……)

③ We may conclude that…或 We come to the conclusion that…(我们得出如下结论……)

④ It is generally accepted (believed, held, acknowledged) that…(一般认为……)(用于表示肯定的结论)

⑤ We think (consider, believe, feel) that…(我们认为……)(用于表示留有商量余地的结论)

(2) 关于建议,可用如下表达方式。

① It is advantageous to (do)

② It should be realized (emphasized, stressed, noted, pointed out) that …

③ It is suggested (proposed, recommended, desirable) that …

④ It would be better (helpful, advisable) that…

6. 结尾部分

1) 致谢

为了对曾给予支持与帮助或关心的人表示感谢,在论文之后,作者通常对有关人员致以简短的谢词,可用如下方式:

I am thankful to sb. for sth.

I am grateful to sb. for sth.

I am deeply indebted to sb. for sth.

I would like to thank sb. for sth.

Thanks are due to sb. for sth.

The author wishes to express his sincere appreciation to sb. for sth.

The author wishes to acknowledge sb.

The author wishes to express his gratitude for sth.

2）注释

注释有两种方式：一种为脚注，即将注释放在出现的当页底部；另一种是将全文注释集中在结尾部分。两种注释位置不同，方法一样。注释内容包括：

① 引文出处。注释方式参见"参考文献"。

② 对引文的说明，如作者的见解、解释。

③ 文中提到的人的身份，依次为职称或职务、单位。如：

Professor, Dean of Dept…. University(教授，……大学……系主任)

Chairman, … Company, USA(美国……公司董事长)

④ 本论文是否曾发表过。

3）参考文献

在论文的最后，应列出论文所参考过的主要论著，目的是表示对别人成果的尊重或表示本论文的科学根据，同时也便于读者查阅。参考文献的列法如下：

如果是书籍，应依次写出作者、书名、出版社名称、出版年代、页数。如：

Dailey, C. L. and Wood, F. C., Computation curves for compressible Fluid Problems, *John Wiley & Sons, Inc. New York*, 1949, pp.37-39

如果是论文，应依次写出作者、论文题目、杂志名称、卷次、期次、出版年份、页数。如：

Marrish Joseph G., Turbulence Modeling for Computational Aerodynamics, *AIAA J. Vol-21, No.7*, 1983, PP.941-955

如果是会议的会刊或论文集，则应指出会议举行的时间、地点。如：

Proceedings of the Sixth International Conference on Fracture Dec. 4-10, 1984, New Delhi, India

如果作者不止一人，可列出第一作者，其后加上 et al。如：Wagner, R. S. et al,…

附录 C 常见计算机英语缩写

A

3G,4G,5G	3rd,4th,5th Generation,第三代,第四代,第五代移动通信业务
A/D	Analog/Digital,模拟/数字
AB	Address Bus,地址总线
AC	Alternating Current,交流电
AC-3	Audio Coding-3 (Dolby Digital AC-3),由 Dolby 实验室制定的一个音频标准
ACPI	Advanced Configuration and Power Interface,高级配置和电源管理界面
ACS	Accounting Control System,记账管理系统(Univac 公司)
	Adaptive Computer System,自适应计算机系统
	Advanced Connectivity System,先进布线系统
	Alternating Current Synchronous,交流同步
	Automatic Coding System,自动编码系统
ADCCP	Advanced Data Communication Control Procedure,高级数据通信控制规程
ADO	ActiveX Data Objects,Microsoft 公司的一种新的数据访问模型
ADP	Automatic Data Processing,自动数据处理
ADPCM	Adaptive Differential Pulse Code Modulation,自适应差分脉冲编码调制
ADSL	Asymmetrical Digital Subscriber Loop,非对称数字用户环线
	Analog to Digital Simulation Language,模-数模拟语言
	Asymmetric Digital Subscriber Line,异步数字用户线
ADTS	Audio Data Transport Stream,音频数据传输流
	Automated Data and Telecommunications Service,自动数据和远程通信服务
AGP	Accelerated Graphics Port,加速图形接口
AI	Artificial Intelligence,人工智能
ALI	Asynchronous Line Interface,异步线路接口
	Automatic Location Identification,自动位置识别
ALT	Alternate Key,备用键
	Automatic Line Test,线路自动测试
ALU	Arithmetic and Logic Unit,算术与逻辑部件(运算器,算术逻辑单位)
AMD	Advanced Micro Device,(美国)AMD 公司(主要生产半导体及芯片)
	Air Movement Data,空气运动数据

	Analog Memory Device,模拟存储器件
	Associative Memory Data,相联存储数据
	Associative Memory Device,相联存储器件
AMI	Access Method Interface,存取方法接口
	Alternate Mark Inversion,传号交替变换
	American Micro system,美国微计算机系统
AMR	Audio/Modem Riser,声音、调制解调器插卡
	Arithmetic Mask Register,运算屏蔽寄存器
	Automatic Message Recording,自动信息记录
	Automatic Message Registering,自动信息挂号(登记)
AN	Access Network,接入网
ANSI	American National Standards Institute,美国国家标准协会
AOL	American On-Line,美国在线服务公司
	All On-Line,全部联机
APC	Adaptive Predictive Coding,自适应预测编码(法,技术)
	Angular Position Counter,角位置计数器
	Associative Processor Controller,相联处理机控制器
	Automatic Peripheral Control,自动外(围)设(备)控制(器)
	Automatic Program Control,自动程序控制
API	Application Program Interface,应用程序接口
APM	Advanced Power Management,高级电源管理
	Automatic Predictive Maintenance,自动预测性维护
	Automatic Programming Machine,自动程序设计机
APPC	Advanced Program to Program Communications,先进的程序间通信技术(方法,(子)系统,协议,程序)
	Automatic Power Plant Checker,电源设备自动检验器
ARP	Address Resolution Protocol,地址解析协议
ARPA	Advanced Research Project Agency,高级研究计划局
ARX	Automatic Retransmission eXchange,自动重发交换机
ASCII	American Standard Code for Information Interchange,美国信息交换标准代码
ASP	Active Server Pages,动态服务器页面
ATL	Active Task List,有效任务表
	Active Time List,有效时间表
	Analog Transmission Line,模拟传输线路
	ActiveX Template Library,ActiveX 模板库
	Application Terminal Language,应用终端语言
	Artificial Transmission Line,仿真传输线

	Automatic TeLling,自动出纳
ATM	Asynchronous Transfer Mode,异步传输模式
	Automatic Teller Machine,自动取款[出纳]机
	Auxiliary Tape Memory,辅助磁带存储器
ATX	Advanced Technology eXtended,一种新的 PC 主板架构规范
AVGA	Accelerated Video Graphics Array,加速的视频图形阵列显示卡
AVI	Audio Video Interlaced,Video for Windows 的多媒体文件格式

B

BASIC	Beginner's All-purpose Symbolic Instruction Code,初学者通用符号指令码
BBL	Be Back Later,稍候便回
BBS	Bulletin Board System,电子公告栏系统
BCD	Binary-Coded Decimal,二-十进制码
BDE	Borland Database Engine,Borland 的数据库引擎
BGA	Ball Grid Array,球栅阵列(组件)
BGP	Border Gateway Protocol,边界网关协议
BIOS	Basic Input/Output System,基本输入输出系统
BLOBs	Binary Large Objects,很大的二进制数据块
BNC	Bayonet Nut Connector,同轴电缆接插件
BO	Back Orifice,后门(一种黑客程序)
BPL	Broadband over Power Lines,电力线宽带
Bps	Bytes Per Second,比特每秒
bps	bits per second,位每秒
BRI	Basic Rate Interface,基本速率接口
BTB	Branch Target Buffer,分支目标缓冲器

C

CA	Certificate Authority,数字证书认证中心
CAD	Computer Aided Design,计算机辅助设计
CAE	Computer Aided Engineering,计算机辅助工程
CAI	Computer Aided Instruction,计算机辅助教学
CAM	Computer Aided Manufacturing,计算机辅助制造
CAPP	Computer Aided Process Planning,计算机辅助工艺规划
CASE	Computer Aided Software Engineering,计算机辅助软件工程
CASL	Computer Assembly System Language,计算机汇编系统语言
CAT	Computer Aided Test,计算机辅助测试
CAX	Community Automatic eXchange,公用自动交换(机)

CB	Control Bus,控制总线
CBD	Central Business District,中央商务区
CBE	Computer-Based Education,计算机辅助教育
CCITT	Consultative Committee of International Telegraph and Telephone,国际电报电话咨询委员会
CCW	Channel Command Word,通道命令字
	Channel Control Word,通道控制字
	China Computer World,计算机世界(中国)
	Counter Clock Wise,逆时针方向(的)
CD-DA	Compact Disc Digital Audio,数字音频光盘
CD-I	Compact Disc Interactive,交互式 CD
CDMA	Code Division Multiple Access,码分多路访问(通信)
CD-R	Compact Disc Recordable,一次性可写入光盘
CD-ROM	Compact Disc Read-Only Memory,光盘只读存储器
CD-RW	CD-ReWritable,可重复擦写光盘
CE	Call Entry,调用入口
	Channel End,通道传送结束(通道末端)
	Chip Enable,芯片启动
	Circular Error,循环误差
	Clear Entry,清除输入
	Common Emitter,共射极
	Communication Equipment,通信设备
	Computer Engineer,计算机工程师
CEO	Chief Executive Officer,执行总裁
	Chip Enable Output,芯片启用(使能)输出
CERNet	China Education and Research Computer Network,中国教育和科研计算机网络
CES	Communication Engineering Standard,通信工程标准(日本)
	Computer Education System,计算机教育系统
	Consumer Electronics Show,家用电子产品展览
CGA	Color Graphics Adapter,彩色图形适配器
CGI	Common Gateway Interface,公共网关接口
	Computer Graphics Interface,计算机图形接口
CHINAMDN	China Mobile Directory Number,公用移动数据通信网
CHS	Cylinders, Heads, Sectors,柱面数、磁头数、每柱面扇区数
CIA	Communication Interface Adaptor,通信接口适配器
	Communication Interrupt Analysis,通信中断分析
	Computer Industry Association,计算机工业协会

	Computers Interface Adaptor,计算机接口适配器
	the Central Intelligence Agency of the U.S.,中央情报局
CIMS	Computer-Integrated Manufacturing System,计算机集成制造系统
CIP	Catalog(u)ing In Publication,预编目录,出版过程中编目
	Commercial Instruction Processor,商用指令处理机
	Communication Interrupt Program,通信中断程序
	Console Interface Program,控制台接口程序
CISC	Complex Instruction Set Computer,复杂指令集(系统)计算机
CIT	Cambridge Information Technology,剑桥信息技术
CIX	Commercial Internet Exchange,商业 Internet 交换中心
CJK	China-Japan-Korea,中日韩
CMI	Computer-Management Instruction,计算机管理教学
CMM	Commission for Maritime Meteorology,海洋气象委员会(联合国)
	Computerized Modular Monitoring,计算机化组件监控
	Controllable Memory Module,可控存储组件
CMOS	Complementary Metal Oxide Semiconductor,互补金属氧化物半导体
CNN	Cable News Network,美国有线新闻网络(以提供即时电视新闻报道而闻名)
CNNIC	China Internet Network Information Center,中国互联网络信息中心
COFF	Common Object File Format,公用目标(对象)文件格式
COM	Component Object Model,组件对象模型
	Commercial organizations,[域]商业组织,公司
	Check Operations Manual,检验操作手册
	COMmand,命令
	COMmon compiler,公共编译程序
	COMmon function(Logic Block),公用功能(逻辑块);公共操作
	COMmunications,通信
	Commutator,整流子,换向器,转接器
	COMpiler,编译
COMDEX	Computer Distributor Exph,计算机分销商展览
CORBA	Common Object Request Broker Architecture,公用对象请求代理(调度)程序体系结构
CPU	Central Processing Unit,中央处理器
CRC	Cyclic Redundancy Check,循环冗余校验法[循环冗余核对]
CREF	Cross REFerence,相互参照
CRM	Customer Relationship Management,客户关系管理
CRT	Cathode-Ray Tube,阴极射线管
CDMA	Coding Division Multiplexing Access,码分多路复用,又称码分多址

CSMA/CD	Carrier Sense Multiple Access/Collision Detect,载波监听多路访问/冲突检测
CSP	Communication Scanner Processor,通信扫描处理器
	Coder Sequential Pulse,编码器顺序脉冲
	Commercial Subroutine Package,商用子例行程序包
	Completely-Self-Protected,全自保护
	Control Setting Panel,控制参数设定板
	Control Signal Processor,控制信号处理机
CSP/AD	Cross System Product/Application Development,跨系统(程序)产品/应用软件开发程序
CSP/AE	Cross System Product/Application Execution,跨系统(程序)产品/应用软件执行程序
CSS	Communication SubSystem,通信子系统
	Common Support System,公用支援系统
	Computer Scheduling System,计算机调度系统
	Computer SubSystem,计算机子系统
	Computer System Simulation,计算机系统模拟
	Contact Start Stop,接触起停
CSTNet	China Science and Technology Network,中国科学技术网
CT	Cable TV,有线电视
CTEC	(Microsoft) Certified Technical Education Center,Microsoft 认证高级技术教育中心
CTO	Chief Technical Officer,首席技术执行官
CUA	Channel and Unit Address,通道和单元地址
	Common User Access,公共用户访问(程序)

D

DAC	Digital-to-Analog Converter,数模转换器
DAO	Data Access Object,数据访问对象
DB	Data Base,数据库
	Data Bus,数据总线
	Data Bank,库集,数据库,资料库
	Data Bit,数据位
	Dead Band,静带死区
	DeciBel,分贝
DBA	DataBase Administrator,数据库管理员
DBCC	DataBase Consistency Checker,数据库一致性检查器
DBDD	Data Base Design Description,数据库设计描述

DBMS	Data Base Management System,数据库管理系统
DBS	Data Base System,数据库系统
DC	Direct Current,直流电
	Device Context,设备描述表
DCB	Device Controller Block,设备控制块
DCC	Direct Cable Connection,直接电缆连接
DCD	Dynamic Content Delivery,数据载波检测
DCE	Data Communication Equipment,数据通信设备
DCL	Data Control Language,数据控制语言
DCOM	Distributed Component Object Model,分布式组件对象模型
	Data Center Operations Management,数据中心运行管理
OLE	Object Linking and Embedding,对象连接与嵌入。以前称为网络 OLE
DCR	Document Change Request,文档更改请求
DCT	Discrete Cosine Transform,离散余弦变换
DD	Data Dictionary,数据字典
	Detailed Design Phase,详细设计阶段
	Display Director,同屏显示
DEA	Data Encryption Algorithm,数据加密算法
DDA	Data Decoder Assembly,数据译码器组件
	Data Display Alarm,数据显示报警(装置)
	Demand Deposit Accounting,活期存款记账(程序,软件)
	Digital Differential Analyzer,数字微分分析仪
	Digital Display Alarm,数字显示报警
DDB	Data Display Board,数据显示板
	Descriptive Data Base,描述数据库
	Design Data Book,设计数据手册
	Device Descriptor Block,设备描述块
	Digital Data Buffer,数字数据缓冲器
	Distributed Data Base,分布式数据库
DDC	Display Data Channel,显示数据通道
DDD	Domestic Direct Dial,国内长途直拨
	Detailed Design Document,详细设计文档
DDE	Dynamic Data Exchange,动态数据交换
DDL	Data-Definition-Language,数据定义语言
DDN	Digital Data Net,数字数据网
DDR	Detailed Design Review,详细设计评审
DDRRAM	Double Data Rate,一种随机存取存储器
DDS	Data Dictionary System,数据词典系统

	Data Display System,数据显示系统
	Data Distribution System,数据分配系统
	Data phone Digital System,数据电话数字系统
	Dataphone Digital Service(AT&T),数据电话数字业务(美国电话电报公司)
	Decision Support System,决策支持系统
DDX	Digital Data eXchange,数字数据交换
	Dialog Data eXchange,对话框数据交换
DFT	Diagnostic Function Test,功能诊断测试程序
	Diagnostic Function Tester,诊断功能测试程序
	Discrete Fourier Transform,离散傅里叶变换
	Drive Fitness Tesr,驱动器性能检测
DHTML	Dynamic HTML,动态 HTML
DID	Data Item Description,数据项描述
DII	Dynamic Invocation Interface,动态调用接口
DIME	Dual Independent Map Encoding,双独立地图编码
DIMM	Dual Inline Memory Module,SDRAM 内存条
DIP	Display Information Processor,显示信息处理机(程序)
	Distributed Information Processing,分布式信息处理(技术,方法)
	Draft International standard Proposal,(ISO)国际标准草案
	Dual In-line Package,双列(排)直插式封装
DirectX	Direct eXtension,微软公司对硬件编程的接口,包括 DirectDraw、DirectSound 等
DIS	Digital Immune System,数字免疫系统
DIY	Do It Yourself,自己动手作
DJ	Disc Jockey,(广播电台)流行音乐播音员,流行音乐节目主持人
DLC	Data Line Controller,数据传输线控制器
	Data Link Control,[SNA]数据链路控制
	Data Link Control field,数据链路控制字段
	Digital Logic Circuit,数字逻辑电路
DLL	Dynamic Link Library,动态链接库
DLS	Downloadable Sound,一种音色库存的存放方法
DMA	Direct Memory Access,直接存储器存取
DMI	Desktop Management Interface,桌式管理界面
	Direct Memory Interface,直接存储器接口
DML	Data Manipulation Language,数据操纵语言
DNS	Domain Name System,域名系统
DOC	Data Output Clock,数据输出时钟

	Display Operator Console,显示操作控制台
DOM	Document Object Model,文档对象模型
DPI	Data Pathing Inc.,数据通路公司
DPMI	DOS Protect Modal Interface,DOS 保护模式接口
DPMS	Display Power Manage System,显示器电源管理
DPS	Data Protection System,数据保护系统
	Data Path Switch,数据通路开关
	Data Processing Station,数据处理站
	Data Processing Subsystem,数据处理子系统
	Data Processing System,数据处理系统
	Delayed Printer Simulator,延迟打印机模拟器
	Desk Publishing System,台式出版系统
DRAM	Dynamic Random Access Memory,动态随机存取存储器
DSL	Data Set Label,数据集标号
	Data SubLanguage,数据子语言
	Dialogue Specification Language,对话说明语言
	Digital Simulation Language,数字模拟语言
	Direct Swift Link,直接快连链路
DSO	Data Source Object,数据来源对象
DSP	Digital Signal Processing,数字信号处理
DST	Drive Self Test,驱动器自我测试
DTD	Document Type Definition,文件类型定义
DTS	Data Transformation Services,数据转换服务
DVD	Digital Visual Disc,数字化视频光盘
	Digital Versatile Disc,数字万能光盘
DVI	DeVice Independent,一种电子文件格式

E

EAX	Environmental Audio Extension,环境音效
	Electronic Automatic eXchange,电子自动交换机
EBCDIC	Extended Binary-Coded Decimal Interchange Code,扩展的二-十进制交换码
EC	Electronic Commerce,电子商务
ECC	Error Correction Code,纠错码,如汉明码
	Electron Coupling Control,电子耦合控制
	Electronic Communication Center,电子通信中心
	Electronic Components Conference,电子元件会议
	Electronic Control Circuit,电子控制电路

	Emergency Communi Cation, 应急通信
ECP	Extended Capabilities Port, 扩展并行口
ECTS	European Computer Trade Show, 欧洲计算机交易大展
EDI	Electronic Data Interchange, 电子数据交换
EDO	Error Demodulator Output, 错误解调输出
EDORAM	Extended Data Output RAM, 扩展数据输出随机存取存储器
EEPROM	Electrically Erasable Programmable Read-only, 电可擦除只读存储器
EIA	Electronic Industries Association, 电子工业协会
EIDE	Enhanced Intelligent Device Electronics, 增强型智能设备电子接口
EISA	Extended(Extension) Industry Standard Architecture, 扩展工业标准结构
EMF	Encrypted Magic Folders, 为文件夹加密的工具软件
EPA	Environmental Protection Agency, 美国环保署
	Entry Point Address, 入口点地址
EPP	Enhanced Parallel Port, 增强并行口
ER	Explicit Route, SNA 显式路由
	Exponent Register, 阶寄存器
	Extension Register, 扩充寄存器
	External Register, 外部寄存器
	Entity-Relationship Approach, 实体-关系图
ERM	Employee Relationship Management, 员工关系管理
ERP	Enterprise Resource Planning, 企业资源计划
ESDI	Enhanced Small Device Interface, 加强的小型设备接口
ESS	Electronic Switching System, 电子交换系统
	Electronic Systems Simulator, 电子系统模拟器
EST	Earliest Starting Time, 最早起动时间
	Educational Testing Services, (美国)教育考试服务处
ETSI	European Telecommunications Standards Institute, 欧洲电信标准协会

F

FAQ	Frequently Asked Questions, 常见问题解答
FBI	Federation of British Industries, 英国工业联合会
	File Bus In, 文件总线输入
FCB	File Control Block, 文件控制块
FCC	Federal Communications Commission, (美国)通信委员会
FDDI	Fiber Distributed Data Interface, 光纤分布式数据接口
FDM	Frequency Division Multiplexing, 频分多路复用
FDS	Fast-access Disc Subsystem (CDC), 快速存取磁盘子系统(CDC 公司)
	FielD Separator, 字段分隔符

	File Description Subsystem, 文件描述子系统
	File Description System, 文件描述系统
FIFO	First In, First Out, 先进先出队列
FPM	Facility Power Monitor, 设备功率监控器
	File Protect Memory, 文件保护存储器
FPMRAM	Fast Page Mode RAM, 快速页面模式随机存取存储器
FPU	Floating Point Unit, 浮点运算
FQDN	Fully Qualified Domain Name, 正式域名
FTP	File Transfer Protocol, 文件传送[输]协议
FX	FiXed area, (磁盘上的)固定区

G

GDI	Graphics Device Interface, 图形设备接口
	Generalized Data base Processor, 通用数据库处理机
	Goal Directed Programming, 目标引导的程序设计
GIF	Graphics Interchange Format, 可交换的图像文件
GMA	General Microprocessor Architecture, 通用微处理机体系结构
GMAT	Greenwich mean astronomical time 格林尼治标准天文时
GML	Generalized Markup Language, 通用标记语言
GMR	G-Magnetoresistive Heads, G磁阻磁头
GMT	Greenwich Mean Time, 格林尼治标准时间
	Generalized MultiTasking, 广义多重任务处理
GNP	Gross National Product, 人均国民生产总值
GPP	General Purpose Processor, 通用处理机
GPRS	General Packet Radio Service, 通用分组无线业务
GPS	Global Position System, 全球定位系统
	General Problem Solver, 通用问题求解程序, 通用解题程序
	General Program Solution, 通用程序解
	General Programming Subsystem, 通用程序设计子系统
	General Purpose Simulation, 通用模拟
	General Purpose Simulator, 通用模拟程序
GSM	Global System(Standard) for Mobile Communications, 全球数字移动电话系统
	Generalized Sequential Machine, 通用序列计算机
	Graphics System Module, 图(形)系统模块
GUI	Graphics User Interface, 图形用户界面

H

HSD	Hardware Safety Device,硬件安全装置
HTML	Hypertext Markup Language,超文本标记语言
HTTP	Hypertext Transfer Protocol,超文本传输协议
HWM	HighWay Memory,总线存储器
	HardWare Monitoring,硬件监视

I

I/O	Input/Output,输入输出
IAAF	International Amateur Athletic Federation,国际业余田径联合会
IAB	Internet Architectrue Board Internet,架构委员会
IAC	Information Analysis Center,信息分析中心
	Institute for Advanced Computation,高级计算技术研究所
	International Advisory Committee,国际咨询委员会
	International Association for Cybernetics,国际控制论协会
	Internationl Algebaic Language,国际代数语言
IC	InterChange,高速公路转换出入口
	Integrate Circuit,集成电路
	Instruction Counter,指令计数器
ICC	International Compute Centre,国际计算中心
iCOMP	Intel Comparative Microprocessor Performance,微处理器效率标准。由Intel 公司提出,用于对 Intel CPU 的相对性能做出直观比较的指标
ICP	Internet Content Provider,Internet 内容提供商
	Initial Connection Protocol,初始连接协议
ICQ	I Seek You,网络寻呼机
ID	IDentification,标识(符号),(身份)识别
	IDentification characters,标识字符,识别符
	IDentifier,标识符
IDC	Internet Data Center,Internet 数据中心
	Intelligent Disk Controller,智能磁盘控制器
	International Data Corporation,国际数据公司
	International Development Center,国际开发中心
IDDD	International Direct Distance Dialing,国际直接长途拨号
IDE	Integrated Drive Electronics,集成电路设备
	Integrated Development Environment,集成开发环境
	Intelligent Device Electronics,智能设备电子接口
IDL	Instruction Definition Language,指令定义语言

	Interface Description(Definition) Language,接口描述(定义)语言
	Intermediate Data Description Language,中间数据描述语言
IE	Internet Explorer,Internet 浏览器
IEEE	Institute of Electrical and Electronics Engineers,电气和电子工程师学会
IETF	Internet Engineering Task Force,Internet 工程任务组
IIS	Internet Information Server,Internet 信息服务器
IMAP	Internet Message Access Protocol,Internet 消息访问协议
IMEI	International Mobile Equipment Identity,国际移动设备识别码
IMF	International Monetary Fund,国际货币基金组织
IP	Internet Protocol,网际协议
	Identification of Position,位置标识
	Identification Point,标识点
	Image Processor,图像处理机
	Imagery Processing,图像处理
	Impact Printer,击打式打印机
	Index of Performance,性能指标
	Information Processor,信息[情报]处理机
IRC	Internet Relay Chatting,在线聊天系统
	Information Retrieval Center,情报检索中心
IRQ	Interrupt Request,中断请求
IS	Impact Switch,碰撞式开关
	Impulse Sender,脉冲发送器
	Incoming Sender,输入发送器
	Indexed Sequential,索引顺序(方式)
	Information Science,信息科学
	Information Separator,信息分隔符
	Information Separator character,信息分隔符
ISA	Industry Standard Architecture,工业标准结构
ISAM	Index(ed) Sequential Access Method,索引顺序存取方法
	Integrated Switching And Multiplexing,集中转接和多路复用(技术)
ISAPI	Internet Server API(Internet Server Application Programming Interface),为 Microsoft 所提的 Internet server 的 API
ISDN	Integrated Services Digital Network,综合业务数字网
ISO	International Standards Organization,国际标准化组织
ISP	Internet Service Provider,Internet 服务提供者
ISV	Input Signal Voltage,输入信号电压
	Intensified Silicon Vidicon,增强型硅(靶)视像管
IT	Information Technology,信息技术

J

J2EE	Java 2, Enterprise Edition, 一种应用开发工具
JPEG	Joint Photographic Experts Group, 联合图像专家组（JPEG 编码及压缩标准）

K

KDE	Keyboard Data Entry, 键盘数据输入

L

LAN	Local Area Network, 局(部区)域网(络)
LBA	Logical Block Address, 逻辑块地址
	Laminar Bus A, 分层总线 A
	Linear Bounded Automaton, 线性有界自动机
	Local Bus Adapter, 局部总线转接器
LCD	Liquid Crystal Display, 液晶显示
LED	Large Electronic Display, 大型电子显示器
	Light Emitting Diode, 发光二极管
LHN	Local Haul Network, 局(部区)域网(络)
LIFO	Later In, First Out, 后进、先出队列
LS	Laser Servo, 激光伺服
LSI	Large Scale Integration, 大规模集成(电路)
	Last Segment Indicator, 最末段指示符
	Logic Status Indicator, 逻辑状态指示器(符)
	Langelier Saturation Index, 朗格利尔饱和指数(化学)
LSIC	Large Scale Integrated Circuit, 大规模集成电路
LSL	Ladder Static Logic, 梯形静态逻辑
LVDS	Low-Voltage Differential Signaling, 低压差分信号

M

MAN	Metropolitan Area Network, 城域网
MAT	Machine-Aided Translation, 机器辅助翻译
MAU	Media Access Unit, 介质访问单元
MBA	Main Battle Area, 主要战区
	Master of Business Administration, 工商管理硕士
	MassBus Adapter, 总线转接器
MBCS	Multi-Byte Character Systems, 多字节字符系统, 通常也称为 ANSI 字符集

MC	MicroComputer, 微型计算机
MCA	Micro Channel Architecture, 微通道结构
MCDBA	Microsoft Certified Database Administrator, 微软认证数据库管理员
MCI	Multimedia Control Interface, 媒体控制接口
	Magnetic Core Storage, 磁芯存储器
	Maintenance-data Collection System, 维修数据收集系统
	Management Control System, 管理控制系统
	Master Control Set, 主控装置
	Master Control Station, 主控(制)站
	Master Control System, 主控(制)系统
MCSE	Microsoft Certified Systems Engineering, 微软认证系统工程师
MCSP	Microsoft Certified Solution Provider, 微软认证解决方案提供者
MDK	Multimedia Development Kit, 多媒体开发工具包
MFC	Microsoft Foundation Class, MS-Visual C++ 的类库
MIDAS	Multiple Indexing Data Access System, 多索引数据存取系统
MIDI	Musical Instrument Digital Interface, 数码音响
MIME	Multipurpose Internet Mail Extension Protocol, 多用途的网际邮件扩充协议
MIPS	Million Instructions Per Second, 每秒百万条指令
MIS	Management Information Science, 管理信息科学
	Management Information System, 管理信息系统
	Metal-Insulator-Semiconductor, 金属-绝缘体-半导体
	Micro Information System, 微信息系统
	Mobility Information Server, 无线通信服务器
MMC	Main Memory Controller, 主存储器控制器
	Monthly Maintenance Charge, 月维修费用
	Microsoft Management Console, 微软管理控制台
MO	Magneto Optical, 磁光盘
MOD	Message Output Descriptor, 报文(信息)输出描述符
MPC	Multimedia PC, 多媒体计算机
MPEG	Moving Picture Experts Group, 运动图像专家组(数字电视标准,包括三个部分：MPEG-Video、MPEG-Audio 和 MPEG-System)
MPU	MicroProcessing Unit, 微处理器
MR	Magnetoresistive Heads, 磁阻磁头
MSDEV	MS-Develop Studio, 微软提供的软件开发平台,可开发VC++ 和 VJ++ 等系统
MSDN	Microsoft Developper Network, 微软的开发者网络
MSI	Middle Scale Integration, 中规模集成电路

MSL	Mirror Server Link,镜像服务器链路
MSN	Microsoft Service Network,微软网络服务
MSVC	MS-Visual C++,微软出的可视化C++开发平台
MT	Machine Translation,机器翻译
	Magnetic Tape,磁带
	Master Timer,主定时器
	Mean Time,平均时间
	Measured Time,测量时间
	MeasuremenT,测量,量度
	Mechanical Translation,机器翻译
	Mechanical Translation language,机器翻译语言
MTA	Multiple Terminal Access,多媒体终端存取
	Mail Transfer Agent,邮件传送代理
MTBF	Mean Time Between Failures,平均故障间隔时间[平均无故障时间]
MTH	Magnetic Tape Handler,磁带处理机[程序]
MTS	Microsoft Transaction Server,微软事务处理服务器
MTTF	Mean Time To Failure,平均无故障时间
MTU	Maximum Transport Unit,(数据的)最大传输单元
	Magnetic Tape Unit,磁带机
	Memory Transfer Unit,转储装置;存储器传送装置
	Multiplexer and Terminal Unit,多路转换器与终端装置
MUD	Multi-User Dungeons,多用户网络游戏
	Multi-User Domuins,多用户领域
MVP	Multi-Variable Programming,多变量程序设计

N

NaN	Not a Number,非数值(IEEE浮点运算规定其用来表示没有意义的表达式)
NC	Network Card,网卡
NCAS	National Center for Supercomputing,国际超级计算机中心
NCB	Network Control Block,网络控制部件(网络控制程序块)
NCFC	the National Computing and Networking Facility of China,中国国家计算机与网络设施
NCO	Number Controlled Oscillator,数控振荡器
NCP	National Communications information Plan,国家通信情报计划
	Network Control Processor,网络控制处理机
	Network Control Program,网络控制程序
NCP/VS	Network Control Program/Virtual Storage,网路控制程序存储虚拟

NDIS	Network Driver Interface Standard,网络驱动器接口标准
	Non-Destructive Inspection Standard,(日本)非破坏性检查(协会)标准
NetBEUI	NetBIOS Extended User Interface,网络基本输入输出系统扩展用户接口
NIC	Network Information Center,网络信息中心
	Network Interface Card,网络接口卡
	Network Interface Control,网络接口控制
	Normal Input Cause,正常输入条件
NII	National Information Infrastructure,国家信息基础设施(信息高速公路)
NIP	Network Information Center,网络信息中心
NIT	National Applied Information Technology Certificate,全国计算机应用技术证书考试
NNTP	Network News Transfer Protocol,网络新闻传输协议(USE-NET)
NOS	Network Operation System,网络操作系统
NS	NanoSecond,纳(诺)秒
	Network Services,网络服务
	New Series,新系列
	New Signal,新信号
	New System,新体制(系统)
	No Signal,无[非]信号
	NonSequenced,非(顺)序的
	NonStandard,非标(准),规格外
	Not Signed,无签名
NSF	the National Science Foundation,国家科学基金会
NSP	Network Service Protocol,网络服务协议
	NonStandard Part approval,非标准部件审定
	Numeric SPace character,数字间隔符
	Numeric Subroutine Package,数字子程序包
NT	New Technology,新技术
NTSC	National Television Systems Committee,NTSC制式,全国电视系统委员会制式
NVF	Network Virtual File,网络虚拟文件
NYSE	New York Stock Exchange,纽约证券交易所

O

OA	Office Assistant,办公助手
	Office Automation,办公自动化
OAW	Optically Assisted Winchester,光学辅助温氏技术
OBJ	Object(objection),物体

OCX	OLE Customer Control,对象链接与嵌入控制程序	
ODBC	Open Database Connectivity,开放式数据库互接	
ODE	Orbit Data Editor assembly,轨道数据编辑程序汇编	
ODI	Open Data-Link Interface,开放式数据链路接口	
OEM	Original Equipment Manufacturer,原始设备制造厂家	
OH	Off Hook,摘机	
OLE	Object Linking and Embedding,对象链接和嵌入	
OOD	Object Oriented Design,面向对象设计	
OOP	Object Oriented Programming,面向对象的程序设计	
OOPL	Object Oriented Programming Language,面向对象的程序设计语言	
OpenGL	Open Graphics Library,开放式图形界面	
OS	Operating System,操作系统	
OSD	On Screen Display,同屏显示	
OSI/RM	Open System Interconnect Reference Model,开放式系统互联参考模型	
OWL	Object Window Library,Borland C++的类库(对象窗口库)	

P

P2P	Peer To Peer,点对点
PAL	Phase Alternating Line,PAL 制式
PC	Personal Computer,个人计算机
PCB	Page Control Block,页控制块
	Power Circuit Breaker,电源断路开关
	Printed Circuit Block,印制电路板
	Process Communication Block,进程通信块
	Process Control Block,进程控制块
	Program Control Block,程序控制块
PCI	Program Controlled Interruption,程序控制中断
	Peripheral Component Interconnect,PCI 总线
PCL	Printer Control Language,打印机控制语言
PCM	Pulse Code Modulation,脉冲编码调制
	Plug Compatible Mainframe,接插兼容主机
	Plug Compatible Manufactures,接插兼容制造厂
	Plug Compatible Memory,插接兼容存储器
PCMCIA	Personal Computer Memory Card International Association,个人计算机存储卡国际协会
PD	Phase Change Rewritable Optical Disk Drive,相变式可重复擦写光盘驱动器
PDA	Parallel Data Adapter,并行数据适配器

	Personal Digital Assistant, 个人数字助理
	Personal Data Assistance, 个人信息(数据)终端(助手)
	Physical Device Address, 设备物理地址
PDF	Portable Document Format, 可移植文档格式(Adobe)
PERL	Practical Extraction and Report Language, 实用摘录和报告语言
PFD	Position Finding Device, 位置测定位, 测位仪
	Pulse Frequency Diversity, 脉冲频率分集
PGP	Pretty Good Privacy, PGP 加密程序
PHP	Personal Homepage Program, UNIX 的 ASP
PIN	Personal Identification Number, 个人识别号
PLC	Programable Logic Controller, 可编程控制器
PnP	Plug and Play, 即插即用
PPP	Peer-Peer Protocol(Point to Point Protocol), 端对端协议
PRI	Primary Rate Interface, 主速率接口
PS	Packet Switching, 分组[包]交换技术
	PostScript, 页描述语言文件的后缀
PVC	Permanent Virtual Circuits, 永久有效电路
PWS	Personal Web Server, 个人 Web 服务器

Q

QC	Quality Control, 质量控制
QL	QuasiLinear, 准线性的
	Query Language, 查询语言
	Quick Loading, 快速装入

R

RAD	Rapid Application Development, 快速应用程序开发
RAID	Redundant Arrays of Inexpensive Disks, 廉价冗余磁盘阵列
RAM	Random Access Memory, 随机存储器
RARP	Reverse Address Resolution Protocol, 反向地址解析协议
	Royal Academy of Music, (英国)皇家音乐学院
RC	Radix Complement, 补码, 基数补码
	Reader Code, 阅读器编码
	Real Circuit, 实电路
	Receive Common, 共接收
	Record Check, 记录校对
	Record Count, 记录计数
	Regional Center, (电话)分局; 区域中心

	Remote Computing,远程计算
RDRAM	Rambus Dynamic RAM,存储器总线式动态随机存取存储器
RFC	Request For Comment,请求说明,Internet 标准(草案)
	Residual Field Count,剩余区段计数
RFX	Record Field eXchange,记录字段交换
RISC	Reduced(Reduction) Instruction Set Computer,精简指令集(系统)计算机
	Reduced Instruction Set Cycles,精简指令系统周期
RLE	Run-Length Encoding,行[游]程长度编码(法)
ROM	Read Only Memory,只读存储器
RPG	Report Program Generator,报表程序生成程序
RTF	Rich Text Format,Rich-Text 格式
RTL	Resistor Transistor Logic,电阻晶体管逻辑(电路)
	Real-Time Language,实时语言
	Real-Time Link,实时链路
	Resistor Transistor Logic,电阻晶体管逻辑(电路)
	RunTime Library,运行时间库
RTTI	Run Time Type Information,运行时间类型信息

S

SAP	Service Access Point,服务访问点
SAT	Storage Allocation Table,存储分配表
SB	Safe Browsing,安全浏览技术
SB-ADPCM	Sub-Band-Adaptive Differential Pulse Code,子带-自适应差分脉冲编码
SCADA	Supervisory Control And Data Acquisition,监察控制和数据采集
SCSI	Small Computer System Interface,小型计算机系统接口
SDK	Software Development Kit,软件开发工具包
SDLC	Synchronous Data Link Control,同步数据链路控制(协议,此即全称)
	Software Development Life Cycle,软件开发生命周期
SDRAM	Synchronous Dynamic RAM,同步动态随机存取存储器
SGI	Silicon Graphics Inc.,美国图形工作站生产厂商
SGML	Standard for General Markup Language,通用标记语言标准
	Standard Generalized Markup Language,标准通用标记语言
SIMD	Single Instruction Multiple Data stream,单指令多数据流
SIMM	Single In-line Memory Module,单列直插式内存组件
SLIP	Serial Line Interface Protocol,串行线路接口协议
SMDS	Switched Multimegabit Data Service,转换多兆位数据服务
SMP	Simple Management Protocol,(TCP/IP)简单管理协议

	Symmetric Multiple Processor,对称多处理器
SMTP	Simple Mail Transfer Protocol,简单邮件传输协议
SNA	Systems Network Architecture,系统网络体系结构
SNAP	Semantic Network Array Processor,语义网络阵列处理机
	System for NAtural Programming,自然程序设计系统
	System Network Activity Program,系统网络动作程序
	System Network Architecture Program,系统网络结构程序
SPD	Software Product Description,软件产品说明书
SPEC	Systems Performance Evaluation Committee (System Performance Evaluation Cooperative Consortium),系统性能评定委员会
SQL	Structured Query Language,结构化查询语言
SQL/DS	Structured Query Language/Data System,结构化查询语言/数据系统
SR	Shift Register,移位寄存器
	Special Register,专用寄存器
SRAM	Static Random Access Memory,静态随机存取存储器
SSI	Small Scale Integration,小规模集成电路
SSL	Security Socket Layer,安全套接层
STL	Standard Template Library,标准模板库
STR	STatus Register,状态寄存器
	Synchronous Transmitter/Receiver,同步发送/接收器
SVGA	Super Video Graphics Array,增强型视频图形阵列显示卡

T

TCP	Transmission Control Protocol,传输控制协议
TDM	Time Division Multiplexing,时分多路复用
TRPG	Table Role Play Game,桌上角色扮演游戏
TSR	Temporary Storage Register,暂存寄存器
	Terminate and Stay Resident,驻留内存;终止并驻留,终止后驻留内存(程序)

U

UCP	Uninterruptible Computer Power,计算机用不间断电源,无中断的计算机电源
	Unit Construction Principle,单元结构原理
UCS	Universal Code Set,通用多八位编码字符集
UDP	User Datagram Protocol,用户数据报协议
UI	Unnumbered Information(frame),未编号信息(帧)
	User Interface,用户界面[接口]

	Ultra ATA/66,一种硬盘接口技术
UMA	Upper Memory Area,上端内存(内存中 640KB 到 1MB 的区域)
UML	Unified Modeling Language,统一建模语言
UNC	Universal Naming Conversion,通用命名标准
	Universal Navigation Computer,通用导航计算机
UNIX	Uniplexed Information and Computing System(CS 用 X 代替),一种多用户的计算机操作系统
UPS	Uninterruptible Power Supply,不间断电源
URL	Uniform Resource Locators,统一资源定位器
USB	Universal Serial Bus,(Intel 公司开发的)通用串行总线

V

VBX	Visual Basic Custom Control,可视化 BASIC 语言程序
VCD	Video Compact Disc,视频高密光盘
VCL	Visual Component Library,可视化控件库
VDM	Varian Data Machines,Varian 数据机
	Video Display Module,视频显示组件[模块]
	Vienna Development Method,维也纳研制法
VESA	Video Electronics Standards Association,视频电子标准协会
VGA	Video Graphics Array,视频图形阵列显示卡
VIP	Very Important Person,大人物,要人
VLB	VESA Local Bus,由 VESA 开发的一种局部总线
VLSI	Very Large Scale Integration,超大规模集成电路
VM	Virtual Machine,虚拟机
VML	Virtual Memory Level,虚拟存储级
VMT	Virtual Method Table,虚拟方法表
VOD	Video-On-Demand,视频点播
VOX	Voice-Operated Control,声控器
	Voice-Operated Relay Circuit,声控继电器电路
	Voice-Operated Transmission,声控传输
VRML	Virtual Reality Moduling Language,虚拟现实造型语言
VSS	Visual SourceSafe,数据库格式
VTM	Vacuum Tube Modulator,真空管调制器
	Vocal Tract Models,声道[发声系统]模型

W

W3C	World Wide Web Consortium,国际共同认可的非盈利组织,其宗旨为尽力提升与维护万维网

WAIS	Wide Area Information Server,广域信息服务器
WAN	Wide Area Net,广域网
WAP	Wireless Application Protocol,无线应用协议
WB	World Bank,世界银行
	Wide Band,宽(频)带
	Wirte Buffer,写入缓冲区(器)
	WideBand Coupler,宽带耦合器
WDM	Wave-length Division Multiplexing,波分多路复用
WMA	Windows Media Audio,微软音乐媒体文件
WML	Wireless Markup Language,无线标记语言
WTO	Warsaw Treaty Organization,华约
	Write-To-Operator,写给操作员(的信息)
WYSIWYG	What You See Is What You Get,所见即所得
WWW	World Wide Web,万维网

X

XML	eXtensible Markup Language,可扩充标记语言
XSL	eXtensible Stylesheet Language,可扩充样式语言
XSLT	eXtensible Stylesheet Language Translation,可扩充样式转换语言

附录D 计算机常用词汇

A

access control 访问控制
Access Control List(ACL) 访问控制列表
Active Group 活动组
Active Server Pages(ASP) 一种服务器端的脚本环境,可以用来创建动态Web页或编译Web应用程序
ActiveX Controls ActiveX控件
activity 活动
Address Resolution Protocol(ARP) 地址解析协议
agent 代理
Anonymous File Transfer Protocol(匿名FTP) 匿名文件传输协议
applet 小程序
Application Programming Interface(API) 应用程序接口
array 数组
Asynchronous Transfer Mode(ATM) 异步传输模式
asynchronous transmission 异步传输
authorization 授权

B

bandwidth 带宽
baud 波特
binding 绑定
bits per second(bps) 位/每秒
browser 浏览器

C

cache 高速缓存
call 调用
callback function 回叫功能
catalog agent 目录代理
class 类
customer 客户
client/server architecture 客户端/服务器结构

clustering　群集
colocation　托管
commit　提交
Common Gateway Interface(CGI)　公共网关接口
Component Object Model（COM）　组件对象模型
component　组件
concurrency　并发
control　控件

D

data consistency　数据一致性
Data Encryption Standard(DES)　数据加密标准
datagram　数据报文
data provider　数据提供程序
data source　数据源
Data Source Name(DSN)　数据源名
data source tier　数据源层
data warehouse　数据库仓库
deadlock　死锁
debugger　调试器
default document　默认文档
default gateway　默认网关
design time　设计时间
dial-up　拨号
digital signature　数字签名
Discrete Fourier,Transform(DFT)　离散式傅里叶变换
Discrete cosine,Transform(DCT)　离散余弦变换
domain name　域名
Domain Name System(DNS)　域名系统
dual interface　双重接口
dynamic binding　动态绑定
Dynamic Host Configuration Protocol(DHCP)　动态主机配置协议
dynamic HTML(DHTML)　动态 HTML
Dynamic-Link Library(DLL)　动态链接库
dynamic page　动态页

E

e-commerce　电子商务

e-mail　电子邮件
early binding　前期捆绑
encapsulate　封装
encryption　加密
Ethernet　以太网
event　事件
event handler　事件句柄
event handling　事件处理
event method　事件方法
exception　异常
executable program　可执行程序
expires header　过期头
extended partition　扩展分区
eXtensible Markup Language(XML)　可扩展标记语言
eXtensible Stylesheet Language(XSL)　可扩展样式表语言

F

failback　故障恢复
failover　故障转移
fat server　胖服务器
Fast Fourier Transform(FFT)　快速傅里叶变换
fault tolerance　容错
File Allocation Table(FAT)　文件分配表
file system(FAT)　文件系统
file name extension mapping　文件名扩展名映射
file space　文件空间
File Transfer Protocol(FTP)　文件传输协议
filter　筛选器
filtering, host name　过滤,主机名
filtering, IP address　过滤,IP 地址
firewall　防火墙
footer　页脚
form　表单
frame　帧
Frequently Asked Questions　常见问题
friendly name　好记的名称
FrontPage Server Extensions　FrontPage 服务器扩展
failure to access code page font file　访问代码页字库文件失败

failure writing to device　写入设备失败

Fatal error：can not allocate memory for DOS　不能为DOS分配内存

FDD controller failure　软盘控制器故障

file access denied　拒绝文件访问

file allocation table bad　文件分配表损坏

File already exists. Overwrite?　文件已存在,需覆盖吗?

file can not be converted　文件不能被转换

file cannot be copied onto itself　文件不能自我复制

file cannot to EXE2BIN　文件不能转换成EXE2BIN

file creation error　文件建立错误

file damaged, cannot recover　文件被破坏,不能恢复

file environment error　文件环境错误

file error　文件错误

file exists cross-device link　文件存在交叉链接

file larger than 64KB, cannot load　文件大于64KB,不能装载

file name must be specified　必须指定文件名

file read aborted　文件读取失败

file sharing conflict　文件共享冲突

file were backed up　文件备份

first diskette bad or incompatible　第一张磁盘损坏或不兼容

Fourier transform　傅里叶变换

G

gateway　网关

Globally Unique Identifier(GUID)　全球唯一标识符

Graphical User Interface(GUI)　图形用户界面

Graphics Interchange Format(GIF)　图形交换格式

H

heap(Windows heap)　堆栈(Windows堆栈)

home directory　主目录

home page　主页

host　主机

host name　主机名

hyperlink　超链接

hypertext　超文本

Hypertext Markup Language(HTML)　超文本标记语言

I

icon 图像
ideal noise diode 理想噪声二极管
identification 识别
identification code 识别码
identification, exchange(XID) 交换识别
identifier 识别码
identifier, service access point(SAPI) 服务存取端识别码
identifier, terminal endpoint(TEI) 终端末端识别码
identifier, terminal equipment(TEI) 终端设备识别码
identifier, virtual circuit(VCI) 虚拟电路识别码
idle 闲置
idle character 闲置字符
if-then-else 若……则……否则
illegal character 非法字符
illegal code 非法代码
illegal instruction 非法指令
illumination 照明
image 图像
image analysis 图像分析,析像
image generator 图像生成器
image map 图像图
image mapping 图像映射,图像变换
image processing system 影像处理系统
image sensor 影像感应器
image transfer 影像转移
image, bit-mapped raster 位映射光栅影像
image, erect 正立像
image, ghost 重影;鬼影
image, inverse 倒像
image, inverted 倒立像
image, real 实像
image, virtual 虚像
image-acquisition 影像采集
immediate access 快速存取
immediate instruction 立即指令
impact printer 击打式打印机

impact strength   抗冲击强度
impairment   损毁；破坏
impulse   冲量
impulse current   脉冲电流
incremental   增量的
index   索引
index hole   索引孔
index register   索引寄存器
index table   索引表
index, disk   磁盘索引
indexed file   索引文件
indexing, cylinder surface   磁柱面索引法
indicated value   指示值
indication   指示
indices   指数
indirect addressing   间接寻址
Information Sciences Institute(ISI)   ［美］信息科学协会
information interchange   信息交换
information retrieval   信息检索
information storage   信息储存
information theory   信息理论
infrared (IR)   红外线
inhibit   禁止；阻止
initial access latency   初始存取等待时间
initial failure   初始故障
ink-jet printer   喷墨式打印机
input   输入
input device   输入装置
Input/Output(I/O)   输入输出
insert   加入；插入
installation manual   安装说明书
instantiation   例示
instruction   指令
instruction cache   指令高速缓冲存储器
instruction code   指令代码
instruction cycle   指令周期
instruction decode   指令解码
instruction format   指令格式

instruction length 指令长度
instruction path 指令通路
instruction sequence 指令序列
instruction type 指令类型
instruction, computer-aided(CAI) 计算机辅助教学
instruction, cycles-per-(CPI) 每指令周期数
instruction, illegal 非法指令
integer 整数
integrated circuit, very large scale(VLSI) 超大规模集成电路
integrated data processing 综合数据处理
Integrated Services Architecture(ISA) 综合业务结构
Integrated Services Digital Network(ISDN) 综合服务数字网络
integration circuit 积分电路
Intellectual Property Right(IPR) 知识产权,专利
intelligence, artificial(AI) 人工智能
Intelligent Network(IN) 智能网络
Intelligent Peripheral(IP) 智能外设
intelligent terminal 智能型终端机
intelligent test 智能测试,智力测验
intensity 亮度;强度
intensive, central processing unit 中央处理器密集
interaction 交互
intercept 截听
intercept call 截听电话
interchannel interference 通道间干扰
intercom 内部通信
interface repository 接口库
interface, bus 总线接口
interface, computer graphics(CGI) 计算机图像接口
interface, data 数据接口
interface, graphical user(GUI) 图像用户接口
interference 干涉;干扰
interference field strength 干扰场强度
interlace scan 隔行扫描
interlaced 交织的;隔行的
interlaced GIF 隔行 GIF 格式
interlaced scanning 交织扫描;隔行扫描
interpretation 解译

interpreter 解译程序
interpreter, language 语言解译程序
interrupt 中断
interrupt acknowledge 中断确认
interrupt bus 中断总线
inverse 反向；倒数
Inverse Fast Fourier Transform(IFFT) 快速傅里叶逆变换
inverse image 倒像
ion gun 离子枪
ip address IP 地址
isolating switch 切断开关，断路器
isolation boundary 隔离边界

J

jack 插座；塞孔
jamming 干扰；干扰杂音
jamming margin 干扰极限
jitter 抖动
jitter, horizontal 水平抖动
jitter, phase 相位抖动
jitter, timing 时序抖动
jitter, vertical 垂直抖动
joint 结；结点
joystick 摇杆
jump 跳越
jump, indirect 间接跳越
junction 结；连接；结点
junction depth 连接深度
junction diode 结式二极管
Junction Field-Effect Transistor(JFET) 结型场效应管
Junction Isolation(JI) 连接隔离

K

kernel 核心
key value 关键值
key, soft 软键
keyboard 键盘
keying, on-off(OOK) 开关键控

keyword 关键字
know-how 技术知识
knowledge base 知识库

L

label 标记；标志
language interpreter 语言解译程序
language, FORTRAN FORTRAN 语言
language, graphics 图像语言
language, hypertext markup(HTML) 超文本标记语言
language, structured query(SQL) 结构化查询语言
language, virtual reality modeling(VRML) 虚拟现实标记语言
laptop computer 膝上型计算机
laser 激光
latch 锁存
latency, interrupt 中断等待时间
layer, board 电路板层
layer, data link 数据链路层
layer, hardware interface 硬件接口层
layer, link 链路层
layer, network 网络层
layer, network interface(NIL) 网络接口层
layer, session 会话层
layer, transmission 传输层
layout 布局
layout, circuit 电路布局
length, focal (f) 焦长；焦距
length, instruction 指令长度
lens 透镜
library 程序库
license 授权，许可
licensing standard 许可标准
life cycle 生命周期
Light Emitting Diode(LED) 发光二极管
light sensitivity 光敏度
light source 光源
line printer 行式打印机
linear power supply 线性电源供应

Liquid Crystal Display(LCD)　液晶显示器
load, full　全负载
load, half　半负载
location　定位
locator　定位器
logarithm　对数
logic　逻辑
logic AND(LAND)　逻辑与
logic OR(LOR)　逻辑或
logic, complementary　互补逻辑
loop　环路
loop antenna　环形天线
luminous intensity　光度

M

machine address　机器地址
machine code　机器代码
machine language　机器语言
macro　宏指令
magnetic field　磁场
mail server　邮件服务器
mailbox　邮箱
mainframe　大型计算机
maintenance　维修;检修
malfunction　故障;失灵
man-machine interface　人机接口
management, configuration　配置管理
management, distributed data(DDM)　分布式数据管理
management, dynamic storage　动态存储器管理
management, library　程序库管理
management, network　网络管理
map　映射;图;图像;变换
mapping, topological　拓扑绘图
margin　边缘;界限
mark　记号
marker　记号;标记
matrix　矩阵
matrix circuit　矩阵电路

media 媒介;媒体
Media Access Control(MAC) 媒体存取控制
medium 介质;媒体
mega-(M) 百万
mega-cycle 百万周期
megabit(Mb) 百万位
megabyte(MB) 百万字节
megacell 百万储存单元
megapixel 百万像素
memory 存储器
memory available 可用存储器
memory bank 存储器组
memory bus 存储器总线
memory cell 存储器存储单元
memory cell array 存储器存储单元阵列
Memory Control Unit(MCU) 存储器控制单元
memory controller(MEMC) 存储器控制器
memory counter 存储器计数器
memory device 存储器器件
Memory Management Unit(MMU) 存储器管理单元
memory, compact disc read only(CD ROM) 光碟只读存储器
memory, electrically erasable programmable read only(EEPROM) 电气拭除式可编程只读存储器
memory, electrically erasable read only(EEROM) 电气拭除式只读存储器
memory, erasable programmable read only(EPROM) 可拭除式可编程只读存储器
memory, flash 快闪存储器
memory, programmable read only(PROM) 可编程只读存储器
memory, random access(RAM) 随机存取存储器
memory, read only(ROM) 只读存储器
memory, static random access(SRAM) 静态随机存取存储器
memory, video random access(VRAM) 视频随机存取存储器
metabase 元数据库
metadata 元数据
metalanguage 元语言
micro- 微
micro-instruction 微指令
microcode 微编码
microcomputer 微计算机

micron 微米
microsecond 微秒
milli-(m) 毫
Million Instruction Bytes per Second(MIBS) 每秒百万指令字节
Million Instructions per Second(MIPS) 每秒百万条指令
millisecond(ms) 毫秒
mini-computer 小型计算机
mirror effect 镜像效应
mobile phone 移动电话
mode,asynchronous transfer(ATM) 异步传输模式
mode,auto-detect 自动检测模式
mode,auto-zero 自动归零模式
mode,normal 正常状态;正常模式
mode,power-down 省电状态;省电模式
mode,privileged 特许状态;特许模式
mode,protected 保护状态;保护模式
Modem 调制解调器
modulation 调制
modulation,amplitude(AM) 振幅调制
modulation,differential pulse code(DPCM) 差动脉冲编码调制
modulation,frequency(FM) 频率调制
modulation,pulse code(PCM) 脉冲编码调制
modulo(mod) 模数
motherboard 母板
motion 运动
Motion Picture Experts Group 1(MPEG-1) 活动图像专家组规范1
Motion Picture Experts Group 2(MPEG-2) 活动图像专家组规范2
Motion Picture Experts Group 4(MPEG-4) 活动图像专家组规范4
motion estimator 运动估算量,移动估算器
motion vector 运动矢量
motion,circular 圆周运动
motion,curvilinear 曲线运动
motion,period 周期运动
motion,projectile 抛体运动
motion,rectilinear 直线运动
motion,relative 相对运动
mouse 鼠标
Multi-User Dimension(MUD) 多用户空间

multiple inheritance 多重继承
Musical Instrument Digital Interface(MIDI) 乐器数字接口
musical scale 音阶
mute 哑音
muting 噪声抑制
mutual inductance 互感
mutual synchronization 互同步

N

nano-(n) 毫微;纳
National Center for Supercomputing Applications(NCSA) 国家超级计算应用中心,国家计算中心
negate 否定;非
Nested loops 嵌套循环
net 网
network 网络
Network Access Points(NAP) 网络接入点
network access server 网络接入服务器
network adapter 网络适配器
Network Address Translator(NAT) 网络地址转换器,网络地址翻译器
network administrator 网络管理员
network architecture 网络结构
Network Control Protocol(NCP) 网络控制协议
Network File System(NFS) 网络文件系统
network function 网络函数
Network Information Center(NIC) 网络信息中心
Network Interface Card(NIC) 网络接口卡
Network Interface Unit(NIU) 网络接口单元
network layer 网络层
Network Management Unit(NMU) 网络管理单元
network server 网络伺服器
network, digital 数字网络
network, distributed 分布式网络
network, integrated services digital(ISDN) 综合服务数字网络
network, star 星型网络
network, token-ring 令牌环网络
networking node 网络结点
neural network 神经网络

neutral 中性
newsgroup 新闻组
nibble 半字节
nick 刻痕;沟;槽
nine's complement 十进制反码
noise 噪音;噪声
noise, impulse 脉冲噪声
non-conductor 非导体
normal 法线
normal mode 正常状态;正常模式
notation 符号;记号
notch 槽口
notebook computer 笔记本式计算机
notification message 通知消息
null 零
null characters 空字符
number 号码;编号
number portability 数字可移植性,号码可移植性
number, node 网点编号
numerical 数字的;数值的

O

object 目的;物;对象
object distance 目标距离
object file 目标文件
object oriented 目标取向
object program 目标程序
object-oriented language 面向对象语言
Object-Oriented Programming System(OOPS) 面向对象编程系统
octal 八进制
off board 外接;板外
offset address 偏移地址
offset, section 区段偏移
offset, segment 分段偏移
ohm 欧姆
ohmmeter 电阻计
on-board 板上
on-board memory 在板存储器,板上存储器

on-chip 片上;芯片上
on-hook 接通
on-line 连线
On-Line Analytical Processing(OLAP) 联机分析处理
on-line compiler 连线编译器
on-line processing 联机处理
on-line system 在线系统
on-page access 页上存取
one wait-state 一个等候状态
one-loop circuit 单环电路
ones complement 二进制反码
opaque 不透明
opaque body 不透明体
open 开启;开路
open architecture 开放式结构
open source 开放资源
open sources 源代码开放
open system architecture 开放系统结构
open systems 开放系统
operability 操作能力
Operating System(OS) 操作系统
operating voltage 作业电压
operation code(opcode) 运算代码
operational amplifier(opamp) 运算放大器
optical disk 光盘
optical drive 光盘机
optical fiber 光学纤维;光纤
optocoupler 光耦合器
orbit 轨道
oriented, object 面向对象
oriented, process 面向过程
out, first-in first-(FIFO) 先进先出
out, last-in first-(LIFO) 后进先出
Output Input(I/O) 输入输出
overhead time 开销时间
overlap 重叠
overlap angle 重叠角
overload 过载

overwrite 重写;覆写
overwrite error 重写错误;覆写错误

P

package 包装;封装;封装组件;套装软件
packaging density 封装密度
packaging level 封装等级
packed data 数据信息包
packet 信息包
Packet Data Unit(PDU) 分组数据单元,包数据单元
packet header 包报头
packet layer protocol 信息包层协定
packet switching 分组交换
page access 分页存取
page attribute 分页属性
page description language 页描述语言
page fault 分页错误
page frame 分页框架
page interleave 分页交错
page space 页空间
page, graphic 图像分页
page, logical 逻辑分页
page, swapping out 交换出分页
paging segment 分页分段
paging unit 分页单元
paging, virtual 虚拟分页
palette 调色板
palette, color 彩色调色板
palette, true-color 真色调色板
pan 摇全景;摇镜头
panel 平板;面板
panel, front 面板
parallel 平行;并行;并联
parallel interface 并行接口
parameter 参数
parity 奇偶
parity bit 奇偶位
parity check 奇偶检测

partial 部分
password 口令;密码;通行密码
path 通路
pause 暂停
pause frame 停顿帧,暂停帧
pay-TV 付费电视
payload 有效载荷
peak 峰值
performance 性能
performance optimization 性能优化
performance test 性能测试
period 周期
period motion 周期运动
period,activation 活化期
peripheral 外设设备
peripheral cell 外设存储格
perl programming language perl 程序设计语言
Personal Computer(PC) 个人计算机
Personal Digital Assistant(PDA) 个人数字助理
physical layer protocol 物理层协议
piezoelectric 压电的
pipeline architecture 管线结构
pipeline instruction 管线指令
pipeline operation 管线作业
pipeline processing 管线处理
pit 坑;缺陷;槽
pitch 音调;间距
pitch,dot 点距
pitch,lead 引线间距
pixel 像素
plane 平面;层
plotter 绘图机
polymorphism 多态
port,dual- 双端口
port,parallel 平行端口
port,serial 串行端口
port,serial read 串行读取端口
portability 可移植性

portable　可移接;手提式
Portable Appliance Tester(PAT)　便携式电器检测仪
portable battery　手提式电池
portable electronics　便携式电子设备
positive charge　正电荷
positive electrode　正电极
positive grid　正栅格
power　功率
power amplifier　功率放大器
power control　功率控制
power, magnification　放大功率
preamble transmission　前文传送
precision　精准
precision, double-　双精度
precision, high　高精度
primary address　原地址
primary battery　原电池
print head　打印头
printer　打印机
Printer Command Language(PCL)　打印机指令语言
Printer Description Language(PDL)　打印机描述语言
printer, daisy wheel　菊轮式打印机
printer, dot matrix　点阵式打印机
printer, impact　撞击式打印机
printer, ink-jet　喷墨式打印机
printer, laser　激光打印机
printer, line　行式打印机
printer, screen　丝网印刷机;网印机
printer, solder paste　焊浆打印机
printer, thermal　热感式打印机
printing, ion-flow　离子流打印
printing, near-letter quality　优质字符打印
programmable　可编程
projectile　抛射体
protective layer　保护层
protocol　协议
pulse　脉冲
Pulse Amplitude Modulation(PAM)　脉幅调制

pure tone 纯音
push-pull 推拉式；推挽式
pushbutton 按钮

Q

quad 四元组
quench detector 终止检测器
query 咨询
Query Processing Unit(QPU) 查询处理单元
queue 排列
queue, code 代码排列
quick sort 快速排序

R

Random Access Memory(RAM) 随机存取存储器
Random Access Storage(RAS) 随机存取存储
raster 光栅,扫描栅,网板,屏面
raster graphics processor 光栅图像处理器
reaction 反应
real image 实像
real-time interrupt 实时中断
Real-Time Operating System(RTOS) 实时作业系统
Real-Time Protocol(RTP) 实时协议
recognition, speech 语音辨认
recognition, voice 声音辨认
recognizer, word 字组辨认器
redundancy 冗余度,冗码码,冗余,多余
reflectance 反射率
refresh 更新
refresh cycle 更新周期
region 区域
register, regular 规则寄存器
register, segment 分段寄存器
register, serial 串行寄存器
register, set 寄存器集
register, shift 位移寄存器
relational database 关系数据库
Relational Database Management System(RDBMS) 关系数据库管理系统

remote　远程
remote enable/disable　远程允许/禁止
Remote File Service(RFS)　远程文件服务
Remote login(R-login)　远程登录
Remote Procedure Call(RPC)　远程过程调用
repeater　中继器
repetition rate　重复率
repetitive　重复
replicated directory　复制目录
report　报表
report generation　报表生成
request　要求;请求
request, interrupt(IRQ)　中断要求
reservation　保留;预定
reset　复位
resource request　资源请求
resource, shared-　资源分享
response time　反应时间;响应时间
restore　再存入;还原
resume　恢复
retrace　回扫
retrieve　检索;撷取
return　回送;返回
ring, token　令牌环
rip　分割;割开
robot　机器人
robot arm　机械臂
rocker switch　摇杆式开关
rotator　旋转器
router　路由器
routine　例行程序
routing algorithm　路由算法
routing control　路由选择控制
row　行
rule, generic　通用法则
run time　运行时间
Run-Length Limited(RLL)　游程长度限制

S

sample and hold　采样和保持
sampling　取样
scalar　标量
scalar product　标积
scan　扫描
scan path　扫描路径
scan rate　扫描率
scan, boundary　边缘扫描
scan, double　双扫描
scan, raster　光栅扫描
scanning, interlaced　交织扫描；隔行扫描
scanning, sequential　逐行扫描；顺序扫描
scanning, vertical　垂直扫描
scope　范围
screen　屏幕
screen, fluorescent　荧光幕
search, binary tree　二元树状搜寻
search, comparative　比较式搜寻
search, distributive　分配式搜寻
search, external　外部搜寻
search, internal　内部搜寻
search, linear　线性搜寻
search, sequential　顺序搜寻
second, byte-per-　每秒字节数
second, million instructions per(MIPS)　每秒百万条指令
secondary output　二次输出；次级输出
section　区段
segment　分段
segment address　分段地址
segment offset　分段偏移
segment register　分段寄存器
segment, code　代码分段
segment, paging　分页分段
semiconductor　半导体
sensor　传感器
sensor, image　影像感应器

separator, encoder/decoder data　编码器/解码器数据分隔器
sequence　序列
sequential list　顺序列
sequential mapping　顺序映射
serial　串行
serial adapter　串行配接器
serial bit　串行位
Serial-In Serial-Out　（SISO)串入串出
server　服务器
server, disk　磁盘服务器
server, file　文件服务器
server, network　网络服务器
service, bulletin board(BBS)　电子布告板服务
service, remote file(RFS)　远程文件服务
Services Switching Point(SSP)　业务交换点
Serving GPRS Support Node(SGSN)　GPRS业务支撑点
session　对话
Session Initiation Protocol(SIP)　话路启动协议
session layer　会话层
set　集
set instruction　指令集
set, character　字符集
set-top box　机顶盒
shadow　阴影
shielding　屏蔽
short　短路
short circuit　短路
shutdown　停机
shutter　快门
signal processing　信号处理
simulate　模拟
slot　间隙;磁格;槽口
Small-Scale Integration(SSI)　小规模集成
smart sensor　智能传感器
socket　插座
soft state　软状态
software　软件
software library　软件库

software portability　软件可移植性
software repository　软件存储库
software reusability　软件可复用性
software tool　软件工具
software trigger　软件触发
sort, comparative　比较式排序
sort, distributive　分配式排序
sort, external　外部排序
sort, insertion　插入式排序
sort, interchange　交换式排序
sort, internal　内部排序
sort, merge　合并式排序
sort, partition exchange　划分交换式排序
sort, selection　选择式排序
sound spectrum　声谱
sound speed　声速
source code　源代码
source file　源文件
spanning tree algorithm　生成树算法
spectrum, sound　声谱
speech recognition　语音识别
speech synthesizer　语音合成器
speed　速度
stack　堆栈
stack register　暂存堆栈器
star connection　星状连接
star network　星状网络
star network topology　星状网络拓扑
state diagram　状态图
state, initial　初始状态
state, logic　逻辑状态
state, steady-　稳定状态
state, wait　等候状态
statement　述句;陈述
statement, conditional　条件式述句
station, network　网络站
station, single-attached(SAS)　单配附传讯站
status　状态

status bit 状态位
status byte 状态字节
storage bit 存储位
storage class 存储类别
storage temperature 存储温度
stream 串;流
string 字串
structural analysis 结构分析
structure, tree 树状结构
structure, user 用户结构
sub-addressing 子地址寻址,子寻址
sub-band coding 子带编码
subnet 子网
subphase 子阶段
subroutine 子程序,子例行程序
subset 子集
subsystem 子系统
subsystem bus 子系统总线
superset 超集

T

table 列表;表
table, file allocation(FAT) 文件配置表
table, index 索引表
table, truth 真值表
Tagged Image File Format(TIFF) 已标记图像文件格式(由 Aldus 和 Microsoft 联合开发)
tangent 正切;切线
tape, cassette 卡式带
telecom 电信,电讯,远距离通信
telecommunication service 电信业务;(ISDN)的用户终端业务
teleconferencing 远程会议,电话会议
template 样板;模板
terminal 终端
Terminal Adapter(TA) 终端配接器
terminal mode 终端方式
terminal server 终端服务器
terminal voltage 终端电压

terminal, data   数据终端机
terminal, intelligent   智能型终端机
terminal, keyboard   键盘式终端机
time trigger   时间触发器
timing analysis   定时分析，时序分析
timing constraint   定时限制
timing data   定时数据
timing delay   定时延迟
timing diagram   定时图
token   令牌；权标
Token Bus Controller(TBC)   令牌总线控制器
token bus network   令牌总线网
token ring   令牌环
token ring switch   令牌环交换
tone   色调；音调
tone, dual   双音调
tone, fundamental   基音
tone, pure   纯音
topological inversion   拓扑式转换
topological mapping   拓扑映射
topology   拓扑，拓扑学
touch screen   触按式屏幕
trace   追踪；线迹
transform   变换
transform operation   变换操作
transformation, discrete   离散变换
transition   转换；变换
transition time   转换时间
translation   翻译；转换
Transmission Control Protocol(TCP)   传输控制协议
Transmission Control Protocol/Internet Protocol(TCP/IP)   传输控制/跨网协议
transmission layer   传输层
transmission, digital   数字传输
transmission, serial   串行传输
transmission, synchronous   同步传输
transparent   透明
transport layer   传送层
transpose matrix   转置矩阵

tree 树
tree network topology 树状网拓扑[技术]
tree structure 树状结构
truth table 真值表
tube,cathode ray(CRT) 阴极射线管
tube,vacuum 真空管
type,bus 总线类型

U

undefined 未定义的
unidirection 单向
Uniform Resource Locator(URL) 统一资源定位符,统一资源定位器
Uniform Resource Name(URN) 统一资源名
unit 单位;单元
unit,binary 二进制单元
unit,bus control 总线控制单元
unit,cache 高速缓冲存储器单元
unit,central processing(CPU) 中央处理器
unit,paging 分页单元
universal 通用
universal ADSL 通用 ADSL
universal buffer 通用缓冲器
universal counter 通用计数器
Universal Synchronous Bus(USB) 通用同步总线
upgrade 升级
usability 可用性
user 用户
user interface 用户接口,用户界面
user,end 终端用户
user-defined 用户定义
utility 实用程序
utility bus 公用总线

V

vacant code 空码
vacuum tube 真空管
valid 有效
valid memory address 有效存储器地址
validation 确认

value 数值;价值
vector 矢量,向量
vector product 向量积
vector resolution 向量分解
vector sum 向量和
velocity 速度
verification 验证
verification system 验证系统
vertex 顶点
vertical 垂直的
Very High Scale Integrated Circuit(VHSIC) 超高规模集成电路
Very Large Scale Integrated circuit(VLSI) 超大规模集成电路
video amplifier 视频放大器
video attribute 视频属性
video codec 视频编解码器,视频编译码器
video converter 视频转换器
video database 视频数据库
virtual 虚拟
virtual circuit 虚拟电路
virtual reality 虚拟现实
Virtual Reality Modeling Language(VRML) 虚拟现实建模语言
virus 病毒;计算机病毒
volume 容量;声量
volume control 声量控制

W

wait state 等候状态
wait-state controller 等候状态控制器
wait-state generator 等候状态产生器
warm-up time 预热时间
warning 警告
watchdog 监视器
waterproof 防水的
watertight 不漏水的,不透水的
watt(W) 瓦特
watt-hour(Wh) 瓦特小时
What-You-See-What-You-Get(WYSWYG) 所见即所得
Wide Area Network(WAN) 宽域性网络
wide band 宽频带

wide, byte- 位宽
window 视窗
window size 视窗尺寸,窗口尺寸
window, register 寄存器视窗
wire 金属线
wireless 无线
wireless LAN 无线区域性网络
word 字;字组
word interleaving 字交织,字交错
workstation 工作站
World Wide Web(WWW) 全球互联网
worm 蠕虫程序(在因特网上自动运行的一种程序)
wow 摇晃
wrap 绕线
wrap-around 环绕
write 写入
write access 写入存取
write buffer 写入缓冲器
write cache 写入式高速缓冲存储器
write cycle 写入周期

X

X-ray spectrum X射线谱
x-y recorder x-y记录仪
xDSL 各种数字用户线

Y

yoke 磁头组
Y connection Y连接
Y network Y网络
Y-T display Y-T显示
Y-address Y-地址
Y-connected circuit 星状连接电路

Z

Z modem 在调制解调器链路上传输文件的一种方法,可自动调整传输速率
Z-address Z地址
Z-axis amplifier Z轴放大器

附录 E 工科学生学习英语的基本要求

随着中国改革开放的不断深入,参与国际化事务日益频繁,对外贸易不断深化,我国对大学生英语的学习日益重视。与此同时,随着企业国际交流的频繁,外包行业的兴起,技术的快速发展,社会对工科学生的专业英语提出了迫切需求。

2007年,教育部网站公布了"大学英语课程要求",其中"教学要求"中指出了大学阶段的英语教学要求分为3个层次,即一般要求、较高要求和更高要求。大学英语课程的设计应充分考虑听、说能力培养的要求,并给予足够的学时和学分;应大量使用先进的信息技术,开发和建设各种基于计算机和网络的课程,为学生提供良好的语言学习环境与条件。

1. 一般要求

(1) 听力理解能力:能听懂英语授课,能听懂日常英语谈话和一般性题材的讲座,能听懂语速较慢(每分钟130～150词)的英语广播和电视节目,能掌握其中心大意,抓住要点。能运用基本的听力技巧。

(2) 口语表达能力:能在学习过程中用英语交流,并能就某一主题进行讨论,能就日常话题用英语进行交谈,能经准备后就所熟悉的话题作简短发言,表达比较清楚,语音、语调基本正确,能在交谈中使用基本的会话策略。

(3) 阅读理解能力:能基本读懂一般性题材的英文文章,阅读速度达到每分钟70词。在快速阅读篇幅较长、难度略低的材料时,阅读速度达到每分钟100词。能就阅读材料进行略读和寻读。能借助词典阅读本专业的英语教材和题材熟悉的英文报刊文章,掌握中心大意,理解主要事实和有关细节。能读懂工作、生活中常见的应用文体的材料,能在阅读中使用有效的阅读方法。

(4) 书面表达能力:能完成一般性写作任务,能描述个人经历、观感、情感和发生的事件等,能写常见的应用文,能在半小时内就一般性话题或提纲写出不少于120词的短文,内容基本完整,中心思想明确,用词恰当,语意连贯,能掌握基本的写作技能。

(5) 翻译能力:能借助词典对题材熟悉的文章进行英汉互译,英汉译速为每小时约300个英语单词,汉英译速为每小时约250个汉字。译文基本准确,无重大的理解和语言表达错误。

(6) 推荐词汇量:掌握的词汇量应达到约4795个单词和700个词组(含中学应掌握的词汇),其中约2000个单词为积极词汇,即要求学生能够在认知的基础上在口头和书面表达两个方面熟练运用的词汇。

2. 较高要求

(1) 听力理解能力:能听懂英语谈话和讲座,能基本听懂题材熟悉、篇幅较长的英语广播和电视节目，语速为每分钟 150～180 词，能掌握其中心大意，抓住要点和相关细节，能基本听懂用英语讲授的专业课程。

(2) 口语表达能力：能用英语就一般性话题进行比较流利的会话，能基本表达个人意见、情感和观点等，能基本陈述事实、理由和描述事件，表达清楚，语音、语调基本正确。

(3) 阅读理解能力：能基本读懂英语国家大众性报刊杂志上一般性题材的文章，阅读速度为每分钟 70～90 词。在快速阅读篇幅较长、难度适中的材料时，阅读速度达到每分钟 120 词。能阅读所学专业的综述性文献，并能正确理解中心大意，抓住主要事实和有关细节。

(4) 书面表达能力：能基本上就一般性的主题表达个人观点，能写所学专业论文的英文摘要，能写所学专业的英语小论文，能描述各种图表，能在半小时内写出不少于 160 词的短文，内容完整，观点明确，条理清楚，语句通顺。

(5) 翻译能力：能摘译所学专业的英语文献资料，能借助词典翻译英语国家大众性报刊上题材熟悉的文章，英汉译速为每小时约 350 个英语单词，汉英译速为每小时约 300 个汉字。译文通顺达意，理解和语言表达错误较少，能使用适当的翻译技巧。

(6) 推荐词汇量：掌握的词汇量应达到约 6395 个单词和 1200 个词组（包括中学和一般要求应该掌握的词汇），其中约 2200 个单词（包括一般要求应该掌握的积极词汇）为积极词汇。

3. 更高要求

(1) 听力理解能力：能基本听懂英语国家的广播电视节目，掌握其中心大意，抓住要点。能听懂英语国家人士正常语速的谈话，能听懂用英语讲授的专业课程和英语讲座。

(2) 口语表达能力：能较为流利、准确地就一般或专业性话题进行对话或讨论，能用简练的语言概括篇幅较长、有一定语言难度的文本或讲话，能在国际会议和专业交流中宣读论文并参加讨论。

(3) 阅读理解能力：能读懂有一定难度的文章，理解其主旨大意及细节，能阅读国外英语报刊杂志上的文章，能比较顺利地阅读所学专业的英语文献和资料。

(4) 书面表达能力：能用英语撰写所学专业的简短的报告和论文，能以书面形式比较自如地表达个人的观点，能在半小时内写出不少于 200 词的说明文或议论文，思想表达清楚，内容丰富，文章结构清晰，逻辑性强。

(5) 翻译能力：能借助词典翻译所学专业的文献资料和英语国家报刊上有一定难度的文章，能翻译介绍中国国情或文化的文章。英汉译速为每小时约 400 个英语单词，汉英译速为每小时约 350 个汉字。译文内容准确，基本无错译、漏译，文字通顺达意，语言表达错误较少。

(6) 推荐词汇量：掌握的词汇量应达到约 7675 个单词和 1870 个词组（包括中学、一般要求和较高要求应该掌握的词汇，但不包括专业词汇），其中约 2360 个单词为积极词汇。

附录 F 部分答案

Unit 1

【Ex1】根据课文内容回答问题：

(1) A general purpose computer has four main components: the arithmetic logic unit (ALU), the control unit, the memory, and the input and output devices (collectively termed I/O).

(2) The control unit, ALU, registers, and basic I/O (and often other hardware closely linked with these) are collectively known as a central processing unit (CPU).

(3) The control unit (often called a control system or central controller) manages the computer's various components; it reads and interprets (decodes) the program instructions, transforming them into a series of control signals which activate other parts of the computer.

(4) The ALU is capable of performing two classes of operations: arithmetic and logic.

(5) On a typical personal computer, peripherals include input devices like the keyboard and mouse, and output devices such as the display and printer. Hard disk drives, floppy disk drives and optical disc drives serve as both input and output devices. Computer networking is another form of I/O.

【Ex2】把下列句子翻译成中文：

略。

【Ex3】选择适当的答案填空：

(1) A　　(2) C　　(3) D　　(4) C　　(5) A

Unit 2

【Ex1】根据课文内容回答问题：

(1) C is a computer programming language. That means that you can use C to create lists of instructions for a computer to follow.

(2) Stderr is a special UNIX file which serves as the channel for error messages.

(3) The body of the function is bounded by a set of curly brackets.

(4) They are integer values, floating point values and single character values.

(5) The way your program remembers things is by using variables.

【Ex2】把下列句子翻译成中文：

(1) 严谨的程序确保了边界条件是正确的。

(2) 括号里包含了参数和它们的类型,用逗号分隔。

(3) 该示例函数可以被另一函数的一行语句调用,该行语句看过去像这样。

(4) 函数体用一对大括号分开。

（5）这是 printf 语句的变量，fprintf 输出语句将它的输出发送到一个文件。

（6）标准输出文件是一个作为错误信息通道的特殊 UNIX 文件。

（7）即使该程序的正常输出被重定向到一个文件或打印机，发送到标准输出文件的消息也将出现在屏幕上。

（8）这句是调用 Power 函数，并将返回值赋值给 result 变量。

【Ex3】选择适当的答案填空：

(1) B　　(2) A　　(3) A　　(4) A　　(5) C

Unit 3

【Ex1】根据课文内容回答问题：

（1）Discrete Mathematics is the general term for several branches of mathematics, which is based on the study of mathematical structures that are fundamentally discrete rather than continuous.

（2）Research in discrete mathematics increased in the latter half of the twentieth century partly due to the development of digital computers which operate in discrete steps and store data in discrete bits.

（3）The topics in Discrete Mathematic are Theoretical computer science, Mathematical logic, Set theory, Graph theory, Operation research, Topology.

（4）Logical formulas are discrete structures, as are proofs, which form finite trees or, more generally, directed acyclic graphstructures.

（5）Operations research techniques include linear programming and other areas of optimization, queuing theory, scheduling theory, network theory. Operations research also includes continuous topics such as continuous-time Markov process, continuous-time martingales, process optimization, and continuous and hybrid control theory.

【Ex2】把下列句子翻译成中文：

（1）图论是一门有许多现代应用的古老学科。

（2）集合的元素之间的关系被表示成一种结构，这种结构叫作关系。

（3）离散数学的不少内容是研究用于表示离散对象的离散结构。

（4）离散数学是通向所有数学学科高级课程的必经之路。

（5）计算机芯片主要负责执行指令。

（6）磁带必须顺序读写，不能随机读写。

（7）可以通过点击选定文本之外的任何地方或按任一方向键取消选定。

（8）单个的 USB 接口比许多并口速度快，它还可以无需端接器而链接许多设备。

【Ex3】选择适当的答案填空：

(1) D　　(2) B　　(3) C　　(4) D、A

Unit 4

【Ex1】根据课文内容回答问题：

(1) A software process is a set of activities that leads to the production of a software product.

(2) They are Software specification, Software design and implementation, Software validation and Software evolution.

(3) Software processes can be improved by process standardisation where the diversity in software processes across an organisation is reduced.

(4) Because of the costs of produdng and approving documents, iterations are costly and involve significant rework. Therefore, after a small number of iterations, it is nonnal to freeze parts of the development.

(5) Errors and omissions in the original software requirements are discovered. Program and design errors, emerge and the need for new functionality is identified. The system must therefore evolve to remain useful.

【Ex2】把下列句子翻译为中文：

(1) 一个软件过程是一组引发软件产品生产的活动。

(2) 由于需要判断和创造力，所以软件过程自动化的尝试只获得了有限的成功。

(3) 计算机辅助软件工程（CASE）工具的有效性受到限制，其中一个原因是软件过程具有极大的差异性。

(4) 对于业务系统，由于需求变更频繁，所以采用一个灵活、敏捷的过程可能更有效。

(5) 软件过程模型是软件过程的抽象表示法。

(6) 系统开发过程主要是把这些组件集成到新系统中，而非从头开发。

(7) 在这个阶段，软件设计是作为一组程序或程序单元实现的。

(8) 瀑布模型的优势在于它在每个阶段都生成文档，而且它与其他工程过程模型一致。

【Ex3】选择适当的答案填空：

(1) B　　　(2) C　　　(3) D　　　(4) B　　　(5) C

Unit 5

【Ex1】根据课文内容填空：

(1) system　(2) manipulates　(3) capability　(4) conflict　(5) execute

【Ex2】把下列句子翻译成中文：

(1) MySQL 数据库系统使用的是客户端/服务器架构，这种架构以服务器为中心。服务器是真正操作数据库的程序。

(2) 服务器是真正操作数据库的程序，客户端不直接对其操作。

(3) MySQL 是与生俱来的网络数据库系统，所以，客户端能和运行在本地机上的或者运行在其他地方的服务器通信。

(4) 这样，一个客户端是一个包括在 MySQL 分布的 MySQL 程序。

(5) MySQL 同时也能使用在非交互方式。例如，从文件或者其他程序阅读语句。

(6) 这使你能从脚本内或者时钟守护作业里，或者连接到其他应用程序使用

MySQL。

（7）另一个流行接口是 phpMyAdmin，它能让用户通过网络浏览器访问 MySQL。

（8）在这种情况下，一些原则可能不同，如结束 SQL 语句的方式。

【Ex3】选择适当的答案填空：

（1）A　　　　（2）A　　　　（3）B　　　　（4）D　　　　（5）C

Unit 6

【Ex1】根据课文内容填空：

（1）unmanned　（2）probes　（3）computexized　（4）fight　（5）lives

【Ex2】把下列句子翻译成中文：

（1）一个嵌入式系统就是一个计算机硬件和软件的集合体，也许还包括其他一些机械部件，它是为完成某种特定的功能而设计的。

（2）在整个80年代，嵌入式系统静悄悄地统治着微处理器时代，并把微处理器带入了我们个人和职业生活的每一个角落。

（3）例如，如果一个实时系统是飞机飞行控制系统的一部分，那么一个延时计算就可能使乘客和机组人员的生命受到威胁。

（4）硬件从一组电连接上读取数字信号，然后转换成模拟信号，传到相连的电话线上。

（5）如果你很幸运，硬件所带的文档中会有你所需要的一整套方框图。

（6）如果你从这个角度想，很快就会发现处理器有很多。

（7）外设不仅是简单的存储传给它的数据，而是把它翻译成一条命令或者翻译成以某种方式处理的数据。

（8）已经有很多具有巨大市场潜力的新的嵌入式设备了：可以被中央计算机控制的调光器和恒温器，当有小孩或矮个子的人在时，智能气囊不会充气，掌上电子记事簿和个人数字助理（PDA）、数码照相机和仪表导航系统。

【Ex3】选择适当的答案填空：

（1）B　　　　（2）A　　　　（3）D　　　　（4）B　　　　（5）C

Unit 7

【Ex1】根据课文内容回答问题：

（1）The Internet Protocol Suite is the set of communications protocols used for the Internet and other similar networks.

（2）The Transmission Control Protocol (TCP) and the Internet Protocol (IP).

（3）The TCP/IP model consists of four layers. From lowest to highest, these are the Link Layer, the Internet Layer, the Transport Layer, and the Application Layer.

（4）In the early 1970s.

（5）An application uses a set of protocols to send its data down the layers, being further encapsulated at each level.

【Ex2】把下列句子翻译成中文：

（1）因特网协议组是一组实现支持因特网和其他类似网络运行的网络传输协议。

（2）更上层越逻辑化地接近使用者，用于处理比较抽象的数据，它依赖低层协议将数据翻译成最终能在物理上传输的形式。

（3）因特网协议起源于1970年早期的国防高级研究项目机构（DARPA）进行的研究和开发。

（4）1978—1983年，几种TCP/IP原型在多个研究中心进行开发。

（5）TCP/IP协议组通过封装提供抽象协议和服务。

（6）这一个模型并不希望成为一个严格的参考模型，并要求新协议以它为标准。

（7）在当今使用的大部分操作系统中，包括所有的用户系统，都含有TCP/IP实现。

（8）大部分的IP实现，程序员可通过使用套接字（使用的也可以是其他协议）和合适的应用接口（API）功能完成。

【Ex3】选择适当的答案填空：

(1) B   (2) C   (3) D   (4) C   (5) A

Unit 8

【Ex1】根据课文内容回答问题：

（1）Array: Quick insertion, very fast access if index known while Slowing search, slowing deletion, fixed size.

（2）Binary tree: Quicking search, insertion, deletion (if tree remains balanced) while Deleting algorithm is complex.

（3）Algorithm is a method to make search, insert, delete a item more effectively.

（4）Procedural programs are organized by dividing the code into functions, so when several functions needed to access the same data, it is easy to bring troubles.

（5）Every entity in our real word can be seems as an object. Objects of the same type can be called a class when you need to look up tens of thousands of items in less than a second.

【Ex2】把下列句子翻译成中文：

（1）基于Java的数据结构与算法是把数据处理成你想要的方式存储的最实用的方式。

（2）迭代的观点对设计具体的算法非常重要。

（3）多组函数可以形成更大的单元，被称为模块或文件。

（4）程序中的对象不仅与现实中的对象对应得更加准确，而且还能解决过程处理模型中全局变量存在的隐患。

（5）程序中的全局变量可以不受限制地被任意函数访问。

（6）链表可能是继数组之后用得最广泛的数据存储结构。

（7）链表是一种灵活的存储机制，适用于多种通用数据库。

（8）对某些哈希表来说，如果表满了，性能将严重下降。

【Ex3】选择适当的答案填空：

(1) C　　　(2) B　　　(3) A　　　(4) A　　　(5) B

Unit 9

【Ex1】根据课文内容填空：
(1) prohibits　(2) restriction　(3) generates　(4) enthusiast　(5) appropriate

【Ex2】把下列句子翻译成中文：
(1) MSDN 是微软生产的一部分(产品)，它负责处理公司开发人员与测试人员的关系，因为硬件开发人员感兴趣的是操作系统，而开发人员面对的是各个操作系统的平台，开发人员利用 API 及微软的应用程序脚本语言。

(2) 当鲍勃离开 MSDN 时，丹尼斯克雷恩博士接手了他的工作任务，并对所负责的栏目增加了一些幽默元素。

(3) 组件是 IComponent 接口的默认实现，并作为在公共语言运行库的所有组件的基类。

(4) 在这种情况下，你可以从 AsyncCompletedEventArgs 类派生自己的类，在派生类中增加私有实例变量和相应的只读公有属性。

(5) System. ComponentModel 命名空间提供了一些用于实现运行时和设计时的组件和控件行为的类。

(6) MSDNAA 账户不是 MSDN 的账户，不能用来访问 MSDN 网站订阅者的内容或下载部分。

(7) 总的来说，它是一个很好的一平台，通过该平台可以正确处理所有的 MSDN 任务。

(8) 如果你添加了 System. ComponentModel. AsyncCompletedEventHandler 委托的事件实例。

【Ex3】选择适当的答案填空：
(1) A　　　(2) D　　　(3) B　　　(4) D　　　(5) C

Unit 10

【Ex1】根据课文内容填空：
(1) complex　(2) yielding　(3) execute　(4) parallel　(5) magnitude

【Ex2】把下列句子翻译成中文：
(1) "优化"这一术语在编译器设计中是指编译器尝试生成比一般代码更高效代码的行为。

(2) 在现代社会，编译器所做的代码优化日益重要和复杂。

(3) 事实上，我们针对编译器优化提出的大多数问题都是不可判定的。

(4) 我们需要对程序的行为完全了解后才能开始，而且要有全面的测试和评估验证我们的直觉。

(5) 优化必须正确，也就是说，要和被编译程序原来的功能一致。

(6) 优化必须提高许多程序的性能。

（7）编译时间必须合理。

（8）如果生成的代码不必正确,那么编写一个能生成快速代码的编译器就是很简单的事情了。

【Ex3】选择适当的答案填空：

(1) A　　　(2) A　　　(3) B　　　(4) C　　　(5) A

Unit 11

【Ex1】根据课文内容回答问题：

(1) An operating system is software, consisting of programs and data that runs on computers and manages the computer hardware and provides common services for efficient execution of various application software.

(2) Including batch processing, input/output interrupt, buffering, multitasking, spooling, runtime libraries, link-loading, and programs for sorting records in files.

(3) The first microcomputers did not have the capacity or need for the elaborate operating systems that had been developed for mainframes and minis.

(4) Microsoft Windows is a family of proprietary operating systems most commonly used on personal computers.

(5) Ken Thompson wrote B, mainly based on BCPL, which he used to write UNIX, based on his experience in the multics project.

【Ex2】把下列句子翻译成中文：

（1）20世纪50年代初,计算机一次只能执行一个程序。

（2）Mac系列操作系统是一款由苹果公司开发、销售,包含其特有知识产权的图形界面操作系统。

（3）20世纪60年代,IBM的OS/360引入一个涵盖了整个产品线的单一操作系统概念。这是System/360成功的关键。

（4）开发成功的微操作系统往往从ROM加载并在监视器上显示。

（5）个人计算机使用的最新版本操作系统是Windows 7,而服务器使用的最新版本操作系统则是Windows Server 2008。

（6）类UNIX家族是一个多元化操作系统,如包括System V、BSD和GNU/Linux等几个主要子类别。

（7）"类UNIX"通常指模拟真正UNIX操作系统的一类操作系统。

（8）符合两种任务要求的典型系统是MINIX,而有些系统(如Singularity)单纯用于研究。

【Ex3】选择适当的答案填空：

(1) C　　　(2) B　　　(3) D　　　(4) D　　　(5) D

附录 G 参 考 译 文

Unit 1 计算机结构

Text A 计算机的功能

1. 中央处理单元和微处理器

一台通用计算机由四大部件构成：算术逻辑单元（ALU）、控制单元、存储器，以及输入输出（I/O）设备。这些部件通常由一组线路构成的总线连接起来。

每个部件都由数以万亿个小电路组成，这些小电路可以被一个电子开关装置进行开关控制。每个电路代表一个比特的信息（二进制数字），使得电路开启时代表一个"1"，关闭时代表一个"0"（正逻辑表示）。电路被设置成逻辑门，使得一个电路的一个或多个状态可以控制其他电路的一个或多个状态。

控制单元、ALU、寄存器以及基本的 I/O（及其紧密耦合的其他硬件）统称为中央处理单元（CPU）。早期的 CPU 由许多独立的元件构成，但 20 世纪 70 年代中期后，CPU 通常构建在一个集成电路上，称为微处理器。

1985 年后，许多处理器设计实现了 MIPS（无互锁装置的流水线微处理器）体系结构的某个版本，并得到广泛使用。MIPS 是指由 MIPS 技术（以前的 MIPS 计算机系统）开发的精简指令集计算机（RISC）指令集体系结构（ISA）。早期的 MIPS 体系结构是 32 位，称为 MIPS32，随后增加了 64 位版本。MIPS32 指令集是与 64 位指令的 MIPS64 指令集同时开发的。MIP32 标准中第一次包含协处理器 0 控制指令。今天，MIP32 指令集是最常见的 MIPS 指令集，与大多数 CPU 兼容。由于其相对简单，MIP32 指令集也是大学课程"计算机结构"中最常见的指令集内容。MIPS32 加法（ADD）操作指令如图 1-1 所示。

控制单元（通常称为控制系统或中央控制器）管理计算机的各种组件。它读取并解释（解码）程序指令，将它们变成一系列激活计算机其他部分的控制信号。先进的计算机控制系统可能会改变一些指令的顺序，以便提高性能。

在所有的 CPU 中，一个关键部件是程序计数器，它是一个特殊存储单元（一个寄存器），用于跟踪下一条指令在存储器中读取的位置。

控制系统的功能如下，注意，这是一个简化的描述，其中一些步骤可能并行处理，或因不同的 CPU 类型以不同的顺序执行。

根据程序计数器指示的单元读取下一条指令的代码。

将数值代码指令解码成其他系统的一组命令或信号。

程序计数器加 1，使其指向下一条指令。

从内存单元（或从输入设备）读取指令所需的任何数据。所需数据的位置通常存储在指令代码中。

提供必要的数据给 ALU 或寄存器。

如果该指令需要 ALU 或指定的硬件完成,则控制硬件执行所请求的操作。

将 ALU 返回的结果写入存储器位置或寄存器,或一个输出设备。

由于该程序计数器(从概念上)仅是另一类存储器单元,它可以通过在 ALU 进行计算而改变。向程序计数器加 100,将导致读取下一条指令是在程序指令再前进 100 的位置。修改程序计数器的指令通常被称为"跳转",允许循环(即由计算机重复进行)并经常与条件指令执行(控制流的两个例子)。

值得注意的是,控制单元经过处理的指令序列引起的系列操作本身就像一个短小的计算机程序。事实上,在一些更复杂的 CPU 设计中,还有另一种更小的计算机称为微程序核,运行微码程序驱动所有这些事件发生。

2. 算术/逻辑单元(ALU)

ALU 能执行两类操作:算术运算和逻辑运算。

特定的 ALU 支持的算术运算集合可能限定为加法和减法,或可能包括乘法或除法或三角函数(如正弦、余弦等)和平方根。有些仅能进行整数(整数)运算,而其他可使用浮点数表示实数,尽管是有限精度。然而,仅能执行最简单操作的任何计算机都可以通过编程,将很复杂的操作分解成它可以执行的简单步骤。因此,任何计算机都可以编程执行任何算术运算,尽管这样做将花费更多的时间,如果它的 ALU 不直接支持这样的操作。一个 ALU 也可比较数值大小并返回布尔真值(True 或 False),这取决于是等于、大于或小于其他数("64 是否大于 65?")。

逻辑运算包括布尔逻辑:AND、OR、XOR 和 NOT。可以创建复杂的条件语句和处理布尔逻辑,这些都有用。

超标量计算机可以包含多个 ALU,使它们可以在同一时间处理多条指令。具有 SIMD 和 MIMD 特性的图形处理器和计算机常常提供可以执行矢量和矩阵算术运算的 ALU。

3. 内存

计算机数据存储通常称为存储或内存,是由计算机组件和记录介质构成的一种技术,用于保存数字数据。如图 1-2 所示的磁芯存储器是一种随机存取计算机内存,主要在 1955—1975 年这 20 年使用。

磁性核心存储器计算机内存占据整个 20 世纪 60 年代,直到它被半导体存储器取代。

可以把计算机存储器当成放置或读取数字的单元列表。每个单元都有一个编号称为"地址",并且可以存储一个数字。可以指令计算机"把数字 123 放到编号为 1357 的单元"或"将单元 1357 里的数加到单元 2468 上,并把答案写进单元 1595"。存储在内存中的信息可能代表实际的任何事情。将字母、数字,甚至计算机指令存入存储器同等容易。由于 CPU 并不区分不同类型的信息内容,所以依靠软件对内存存储的一串数字给出其意义解释。

在几乎所有的现代计算机,每个存储器单元都设置成存储 8 比特组成(称为字节)的二进制数字。每个字节能够代表 256 个不同的数字($2^8 = 256$);无论是 0~255 或者 -128~+127。存储更大的数字,可以使用几个连续的字节(通常使用两个、4 个或 8 个字节)。当需要负数时,它们通常存储成 2 的补码表示法。其他安排也是可能的,但通常看不到,除了专门的应用或历史的上下文。只要能用数字表示,计算机在存储器中就可以存储任何类型的信息。现代计算机拥有数十亿,甚至万亿内存字节。

CPU 包含称为寄存器的一组特殊存储器单元,比主存储读取和写入更迅速。取决于 CPU 的类型,CPU 中通常有 2~100 个寄存器。寄存器用于存储最经常需要访问的数据项,以避免每次需要数据,都访问主存储器。因数据在不断工作,所以减少了要访问主存储器的需求(相对于 ALU 和控制单元,这往往是缓慢的操作),极大地提高了计算机的速度。

计算机内存主要有两种类型:随机存取存储器(或称 RAM)和只读存储器(或称 ROM)。RAM 按照 CPU 的指令随时可读取和写入,但 ROM 是预先装入数据和软件,永远不会改变,所以 CPU 只能从中读出。ROM 经典范例用来存储计算机的初始化启动指令。一般情况下,当计算机电源关闭时,RAM 的内容被删除,但 ROM 永远保留其数据。在 PC 中,ROM 包含一个专门的程序,称为 BIOS,只要计算机开机或重新启动,计算机操作系统程序就将其从硬盘驱动器加载到 RAM。在嵌入式计算机中,经常没有磁盘驱动器,所需的所有软件可以存储在 ROM 中。存储在 ROM 中的软件通常被称为固件,因为它从名称上比软件更像硬件。闪速存储器模糊了 ROM 和 RAM 之间的区别,因为关闭时其数据可以保留,也可以重写。通常,它比常规的 ROM 和 RAM 慢得多,因此,它的使用受到限制,只适用于不需高速访问的场合。

在更复杂的计算机中可以有一个或更多的 RAM 高速缓冲存储器,它们的速度比寄存器慢,但比主存储器快。有这种高速缓存的计算机一般设计成将经常使用的数据自动移动到缓存,往往不需要程序员任何干预。

4. 输入输出(I/O)

I/O 是计算机用来与外界交换信息的通道。提供给计算机的输入或输出设备称为外设。在一个典型的个人计算机中,外设包括输入装置(如键盘和鼠标),以及输出设备(如显示器和打印机)。硬盘驱动器、软盘驱动器和光盘驱动器都被当作输入设备。计算机网络是 I/O 的另一种形式。

通常,I/O 设备是复杂的计算机系统,拥有它们自己专用的 CPU 和存储器。图形处理单元可能包含 50 个或更多个微小的计算机计算需要显示的三维图形。现代台式计算机含有许多小型计算机,可以帮助主 CPU 执行 I/O。

5. 多任务

虽然计算机可看成运行存储在主存储器中的一个巨大程序,但在一些系统中,必须同时运行多个程序并显示出来。这通过多任务处理实现,即让计算机在每个程序之间依次快速切换运行。

这样做的一种方法是,用一个称为中断的特殊信号,它可以周期性地导致计算机停止执行它在那里的指令,而做其他事情。只要记住它被中断之前的执行位置,计算机就可以稍后再返回到该任务。如果几个程序"同时"运行,中断发生器可能会导致每秒几百次中断,每次都使程序进行切换。因为现代计算机执行指令的速度比人类的感觉快几个数量级,它可能显现出许多程序同时运行的现象,即使在任何给定的瞬间永远只有一个程序被执行。多任务的这种方法有时称为"时间共享",因为每个程序依次分配到一个时间"切片"。

在廉价计算机时代来临之前,进行多任务处理的主要用途是让许多人共享同一台计算机。

看似多任务让几个程序之间切换运行,会导致计算机运行更慢,正比于运行程序的数量。然而,大多数程序花费大量的时间等待缓慢的输入输出设备完成它们的任务。如果程序正在等待用户点击鼠标或按下键盘上的一个键,那么它不会花费"时间切片",直到它等待的事件发生。这将释放给其他程序使用,这样,很多程序可能在同一时间运行,而没有不可接受的速度损失发生。

6. 多进程

有些计算机在多处理配置环境设计成分发它们的任务横跨多个 CPU,这一技术曾经只在大型、功能强大的机器,如超级计算机、大型计算机和服务器中使用。多处理器和多核心(在单个集成电路上的多个 CPU)个人计算机和膝上型计算机现在已广泛使用,并且在低端市场也正越来越多地开花结果。

尤其是超级计算机,它常常具有高度不同的独特结构,区别于基本存储程序结构和通用计算机。它们往往拥有成千上万的 CPU、特制的高速互连和专门的计算硬件。这种设计往往只对特定任务有用,同时需要大型程序组织才能成功地、最大限度地利用可用资源。超级计算机通常应用于大型仿真、图形渲染和加密应用,以及其他所谓的"让人为难的并行"任务。

7. 联网和互联网

20 世纪 50 年代开始,计算机已被用于协同共享多个位置之间的信息。美国军方的 SAGE 系统就是一个大型范例,这导致出现一些特殊用途的商业系统,如 Sabre。

20 世纪 70 年代,美国各地研究机构的计算机工程师开始使用电信技术将自己的计算机连接起来。这项工作由 ARPA(现在的 DARPA)资助,而且由它产生的计算机网络被称为 ARPANET。技术发展使得 ARPANET 可以不断扩散和演变。

随着时间的推移,网络传播超越学术和军事机构,并成为著名的互联网。联网的出现使得所涉及计算机的性质和边界需要重新定义。计算机操作系统和应用程序概念进行了修改,使其能力定义包括访问网络上其他计算机的资源,如外设、存储的信息和类似物,作为个人计算机资源的扩展。最初,这些设施主要供高科技环境中的工作人员使用,但 20 世纪 90 年代后,像电子邮件和万维网普及应用,与低价、快速的网络技术(如以太网和 ADSL)发展相结合,计算机联网变得随处可见。事实上,通过网络连接的计算机数量惊人增长。个人计算机定期连接到 Internet 进行通信和接收信息,占非常大的比例。"无线"网络经常使用移动电话网络,意味着网络即使在移动计算环境中,也正在日益普及。

Text B 计算机技术的未来

有段时间,人们一直希望能不断得到最快的计算机,对不对?

当 386 出现后,大家都认为它酷酷的、速度快,每人都需要一台。

突然,应用程序用掉了它的能力,386 变得缓慢,接着 486 就出现了,每人都需要一台。

目前的趋势是，我们充分挖掘硬件性能，成为可用，然后想要更多。有点像"永不满足的钱（欲壑难填）"的说法。

但期望这一趋势将继续下去是存在问题的。

事实上，普通人需要一台计算机做的事就这么多。基本上，计算机作为一个"设备"，能够与我们的感官进行有限的相互作用。这意味着，直到我们看到嗅觉鼠标前，我们基本上需要

全速视频输出：电影解码/演示/产生，游戏。
全速视频输入：视觉识别，电影保存/编码/压缩。
全速音频输出：声音解码/混合，语音合成。
全速音频输入：语音识别，录音/压缩。

这一说法适用于"普通人"，不是像你和我这样的计算机极客，想要运行庞大的仿真程序或做软件开发或 CAD 设计，仍然需要有一个全副功能的家用工作站。我们这类人占少数。

大多数（上面提的）要求，今天的计算机已经可以满足了。其中许多也都可用，只是还没有达到全速（如全屏视频生成、"无瑕"语音识别），并且近几年内购买的现有的计算设备都可以满足。

你有没有注意到，人们不再过多地谈论处理器的速度是多少。如果你到本地计算机商店，他们谈论的多是 RAM、硬盘、屏幕大小、电池和其他功能。几年前，他们谈的都是处理器速度，而现在它变得越来越不重要，不久也许会成为一个脚注，作为一台计算机的规格条件。

那么，接下来会发生什么？

东西会变得更小、更快。

过去，笔记本电脑笨重、速度缓慢且能力微弱。它不能完成一台家用计算机的所有任务。这将不再发生，我的 2 磅超轻薄笔记本电脑（我仅花 150 美元；顺便说一句）基本上能与最先进的家用工作站媲美。可以看到，目前已涌现出功能越来越强大的小型计算机。像 Palm（手持计算辅助设备）这样的设备，只是商业的临时新方向。Palm 仅有真实操作系统的部分功能，也仅有真正计算机的子集功能。很快你就能得到一台和 Palm 大小相同的全功能计算机，而且价格也差不多。你宁愿有一个内置微型浏览器的个人信息管理器，还是宁愿有装有你刚下载最新的 Netscape/Mozilla/Opera 浏览器内含有最新版 Flash/Java/Shockwave 功能的个人信息管理器？还是这样的个人信息管理器，具有提供地址列表的应用程序功能，或可以运行你想要的任何地址列表的应用程序功能，并有玩 Doom（毁灭战士）游戏的功能？

这个论点也适用于数据存储。

每个人都想要他们能得到的最大硬盘。我们不断想出更多的方法满足他们的需求（MP3、视频等），而我们需要不断更新升级。尽管现在，你的硬盘上的存储空间不足是一个非常罕见发生的事情。我曾经不得不压缩文件并将文件保存到软盘上，才能处理一些事情。

现在这些日子我连任何磁盘都没有。在我的计算机上,我通常只写入 CD,以此作为信息的备份方法。

在未来不到 10 年的时间内,你能够将完整的视频存储在硬盘上,就像你今天可以在那里存储音频一样。你会有一个 1000 部的电影播放列表,就像你今天的 MP3 播放器。

几年后,你就可以把它放在你的钥匙串上。之后,你真的需要买一个更大的硬盘吗?在某些时候,你将能够存储更多的视频和音频,你将能够体验你的一生。这时,你就不需要关注下一个驱动器大小了,好吗?

我承认我的笔记本电脑和家里的计算机有很大的区别。主要是键盘、21 英寸的显示器和 CD-RW(可读写光驱)(虽然现在大多数笔记本电脑都有 DVD/CD/RW 等)。

现在让我们来看看未来。

但这样做之前,要考虑过去,回想计算机行业曾发生了什么。

最初,钱主要花在大型机上。个人计算机作为一种奢侈的嗜好而存在,但钱在大型机行业里流动。这一切都与 PC 和大型机相比有多大作用有关。随着性能的提高,服务器开始做很多大型机可以做的事情,而钱仍花在大型机市场上。真正的问题是,有足够程度的计算能力(每美元)实际上是至关重要的。低于这个程度,都会显得太慢,你会想要更快的速度。高于这个能力,在技术和发展市场上只会有用。然后是工作站,它们开始达到计算的能力,并开始替代服务器。然后是个人计算机(PC)。个人计算机市场腾飞,个人计算机开始替换工作站。这个水印特效在电影业中得到了最好的宣传,当时像《泰坦尼克号》这样的电影被广为宣传,是通过使用大量便宜的 PC,而不是工作站制作出来的。现在我们正在转向笔记本电脑,更小笔记本将是下一个,然后是手持设备。自己完成这一推理。

这是我对未来的预测。

你会有一个计算设备。它将是一个处理器,是 xxx MHz,其中 ' xxx ' "足够快"。它也将有固态存储器,有 xxx GB,其中 ' xxx ' "足够大"。这个小家伙将类似手机大小或更小,成本与手机大致相当。

你回家会把它挂接在显示器和键盘上的一个端口上。而在路上,你会把它放在一个带有 Palm 大小触摸显示屏的外壳上,让你访问它的所有信息,甚至听音乐或看视频。然后你可能会将它连上你的相机外壳(或者你的管理者外壳上可能有镜头……),你想拍多少张照片就拍多少张,因为你的存储空间可以容纳 1000 部电影,所以你永远无法用静止照片和家庭视频填满它,也不需与家用设备同步。你甚至可以拥有完全的无线功能,你的照片可以自动传到你的网络相册。当你旅行回家的时候,你可以把这个设备放到你的影院系统里,你可以看电影或从你的旅行镜头中开始观看。你可以在你的设备上加密某种形式的净货币,这样,你也可以用它购买东西。当你开车的时候,可以把它放在车里,听你自己的每一首歌。你可以在自己的车上安装一个 GPS,它可以使用地图查询等服务为你指路。若你的爱人给你发电子邮件,你会用耳机外设和电话软件给她回电话。

但重要的一点是,你不需要得到多个计算设备,也不需要再拥有家庭计算机。你会有一个计算设备,它会像钱包一样伴随你,因为它将取代你的信用卡、你的手机、你的传真机、你的 DVD 播放机、你的立体声、你的浏览器、你的助理器、你的电子邮件等。

然后会有一些博物馆,所有这些东西都摆放在那里,人们可以观看和感受,就像今天我们嘲笑 ENIAC 和真空管一样。

Unit 2　编　程　语　言

Text A　C 语言函数

几乎所有的编程语言都具有与函数等价的功能。你可能已经遇到过称为子程序或过程的别名。

有些语言通过是否有返回值区分函数。C 语言假定每个函数都会返回一个值。若程序员想返回值,通常使用 return 语句实现。如果不需要返回值,调用函数时就不要使用。

这里是一个求双精度数的非负整数次幂的函数,返回计算的结果。

```
double power(double val,unsigned pow)
{       double ret_val = 1.0;
        unsigned i;

        for(i = 0; i < pow; i ++)
                ret_val * = val;

        return(ret_val);
}
```

该函数由一个简单的算法,自乘自身 pow 次。for 循环用来控制乘法的次数,变量 ret_val 用来存储返回值。小心仔细地编程已保证了边界条件是正确的。

让我们看看这个函数的详细信息:

```
double power (double val,unsigned pow)
```

这行表示函数定义的开始。它告诉我们返回值的类型、函数名,以及所使用的函数参数列表。这些参数及其类型都用括号括起来,用逗号分隔。

函数的主体是由一组花括号构成的范围。这里声明的变量将被视为局部的,除非特别声明为 static 或 extern 类型。

```
return(ret_val);
```

到达 return 语句,程序的控制权返回给调用函数。括号内的值是函数的返回值。如果在任何返回值之前到达最终的花括号,那么函数将自动返回,返回值将是毫无意义的。

这个函数例子可以在其他函数中调用,看起来像如下这样:

```
result = power(val,pow);
```

这是调用乘幂函数并将结果赋给变量 result。

下面是一个没有返回值的函数例子。

```
void error_line(int line)
```

```
    {       fprintf (stderr,"Error in input data: line % d\n",line);
    }
```

该定义使用可选的类型 void。这表明没有用到返回值,否则该函数与前面的例子大致相同,不同之处在于该函数没有返回语句。有些 void 类型的函数可以使用 return,而只是强制从函数中提前退出,并且不返回任何值。这有点像使用 break 跳出循环。

这个例子还显示了函数的新特性。

```
fprintf (stderr,"Error in input data: line % d\n",line);
```

这是关于 printf 语句的变体;fprintf 将其输出发送到文件中。在这种情况下,该文件是 stderr。stderr 是一个特殊的 UNIX 文件,其用作错误消息的信道。它通常连接到计算机系统的控制台上,所以这是从你的程序显示错误信息的好方法。即使程序的正常输出已被重定向到文件或打印机上,发送到 stderr 的消息也会出现在屏幕上。

该函数的调用方式如下。

```
error_line(line_number);
```

Text B J2ME 有关资料

1. J2ME 的组织结构如何

传统的计算设备使用相当标准的硬件配置,如显示器、键盘、鼠标以及大容量的内存和永久存储设备。

然而,新一代计算设备缺乏硬件配置的继承性。一些设备不带显示器、永久存储器、键盘或鼠标,并且内存在各小型计算设备相互不兼容。

小型计算设备之间缺乏统一的硬件配置,成为 Java 社区进程项目 JCPP(Java Community Process Program)面临的一个巨大的挑战,该项目负责为小型计算设备的 JVM 和 J2ME 制定标准。

J2ME 必须服务于许多不同种类的小型计算设备,包括视频电话、有线电视数字机顶盒、蜂窝电话和个人数字助理机。JCPP 的挑战是开发出在具有非标准的硬件配置小型计算设备中可以实现的 Java 标准。

JCPP 已经采用了双面的办法解决(满足)小型计算设备的需要。首先,他们定义了每个设备上运行的 Java 运行环境和核心类,这称为配置。配置定义了特定小型计算设备的 Java 虚拟机。有两种配置:一种用于手持式的设备;另一个用于插入式设备。接下来,JCPP 定义小型计算设备类别的概要文件。概要文件包含的类使开发人员能够实现对小型计算设备相关的功能。

2. J2ME 配置

截止到现在,J2ME 有两种配置:有限连接设备配置(CLDC)和连接设备配置(CDC)。CLDC 是为有限内存容量的 16 位或 32 位小型计算设备设计的。

CLDC 设备通常有 160~512KB 可用内存并由电池供电。它们还使用一种不兼容、窄带宽的无线网络连接,并且可以没有用户界面。CLDC 设备使用的 K-Java 虚拟机

（KVM）环境是一个 JVM。CLDC 精简版本,适用装置包括寻呼机、个人数字助理机、蜂窝电话、专用的终端和带有 128～512KB 内存的手持电子设备。CDC 设备使用 32 位架构,至少支持两兆的内存和完整功能的 JVM。CDC 设备包括数字机顶盒、家电、导航系统、销售点终端和智能手机。

3. J2ME 概要文件

概要文件(profile)包含 Java 类,便于一个特定的小型计算设备或一类小型计算设备实现所需要的功能。由于小型计算技术的不断发展,基于此,J2ME 不断定义出新的概要文件。截至目前,已定义了 7 类概要文件,即基础概要文件、游戏概要文件、移动信息设备概要文件、PDA 概要文件、个人概要文件、个人基础概要文件以及 RMI 概要文件。

- 基础概要文件用来参与 CDC 配置,是其他类概要文件的核心,几乎参与所有 CDC 配置,因为其包含核心的 Java 类。
- 游戏概要文件也用于 CDC 配置,包含任何小型计算设备开发的游戏应用程序所需的类。
- 移动信息设备概要文件(MIDP)用来参与 CLDC 配置,包含提供本地存储的类、用户界面和移动计算设备如运行 Palm 操作系统设备的应用程序所需的网络功能。MIDP 用了无线的 Java 应用程序。
- PDA 概要文件(PDAP)用来参与 CLDC 配置,并包括利用复杂资源的类(如个人数字助理机的类)。这些功能包括更好的显示,以及更多的内存,这是与 MIDP 移动设备(如蜂窝手机)所使用的资源相比。
- 个人概要文件用来参与 CDC 配置,需要基础概要文件,还包括用来实现一个复杂用户界面的类。基础概要文件提供核心类,以及个人概要文件提供实现复杂的用户界面的类,这是一个同时能显示多个窗口的用户界面。
- 个人基础概要文件类似于个人概要文件,它用来参与 CDC 配置,需要基础概要文件。然而,个人基础概要文件提供实现一个简单的用户界面的类,这是一个用户界面,能够同时显示一个窗口。
- 将 RMI 概要文件用来参与 CDC 配置,需要基础概要文件,其中包含提供远程方法调用类的基础核心类。

小型计算设备存在许多不断定义的概要文件。JCPP 项目(java.sun.com/aboutjava/communityprocess)的产业群定义了这些概要文件。每个小组确定使用由该行业生产的小型计算设备的标准概要文件。一个 CDC 概要文件针对的是一类小型计算设备,在基础概要文件的基础上由核心 Java 类扩展而成。这些设备特定类包含在一个新的概要文件中,使开发人员能够创建工业级应用。但是,这些基础概要文件仅针对 CDC,而不是在基础概要文件的核心类上针对所有的概要文件都进行扩充。

记住,应用程序仅在 JVM 必要的类或开发者使用的概要文件中能够访问小型计算设备的软件和硬件。

Unit 3 离 散 数 学

Text A 有关离散数学

(1) 离散数学简介

离散数学为几个数学分支的统称,这是基于离散的,而不是连续的数学结构的研究角度的总称。相比实数的"平滑"的特性,离散数学的研究对象如整数、图形和逻辑命题,不以平滑方式变化,而有明显的离散值,因此离散数学不研究"连续数学"的课题,如微积分和分析。离散对象通常可以通过整数枚举。更准确地说,离散数学是处理可数集的数学(与整数的子集具有相同基数的集合,包括有理数,但不是实数集)的分支。然而,"离散数学"这个词还没有精确并为大家接受的定义。事实上,离散数学很少描述研究哪些内容,而更多描述不研究哪些内容,如连续变化的数量及其相关的概念。

20世纪下半叶,由于数字计算机的发展需要研究不连续的操作步骤和存储的数据离散化,因此离散数学得到深入的研究。离散数学的概念和符号应用于研究和描写计算机科学的对象和问题,如计算机算法、编程语言、密码学、自动定理证明和软件开发。结果,计算机成果非常显著。

(2) 离散数学的课题

计算机科学理论包括离散数学与计算相关的领域。它大量借鉴了图论和逻辑学,包括计算机科学理论中的计算数学算法的研究成果。可计算性研究什么原则下可以进行计算,并与逻辑密切相关,而复杂性研究所需要的计算时间。自动机理论和形式语言理论与可计算性密切相关。计算几何算法应用于几何的问题,而计算机图像分析应用于图像的表示。计算机科学理论也包括各种连续计算专题研究。

逻辑学是对有效推理和论证原理的研究,也是对一致性、健全性和完整性的研究。例如,在大多数逻辑系统(但不是在直觉逻辑)中,皮尔斯定律(($p \to q \to p$) $\to p$)是一个定理。对于经典逻辑,它可以很容易地用真值表进行验证。数学证明的研究在逻辑中占有特殊的地位,已经应用于软件的自动定理证明和形式化验证。

逻辑公式是离散结构,就像证明一样,它们形成有限的树,或者更广泛地说,是形式有向无环图形结构(每个推理步骤结合一个或多个前提分支而给出一个单一的结论)。逻辑公式的真值通常形成一个有限集,它一般限于两个值:真值和假值,但逻辑也可以是连续值,如模糊逻辑。也研究无限证明树或无限派生树等概念,如无限逻辑。

集合论是研究集合的数学分支,它们是对象的集合,如｛蓝色、白色、红色｝或所有素数的(无限)集。偏序集和关系集在多个区域具有应用。

在离散数学中,可数集(包括有限集)是主要关注点。集合论作为数学分支的基础,通常以格奥尔格·坎托的工作为标志,区分不同种类的无限集,其动机是三角级数的研究。无限集理论的进一步发展超出了离散数学的范围。事实上,当代的描述集合论的工作广泛运用了传统的连续数学。

图论即图形和网络的研究,往往被认为是组合数学的一部分,但已经变得足够大和清晰,有自己的问题,本身就被视为一门学科,它在数学和科学的所有领域都得到推广应用。

图是离散数学研究的主要对象之一。它们是自然界和人造结构中最普遍的模型之一。它们可以模拟物理、生物和社会制度中的许多类型的关系和过程动态。在计算机科学中,它们代表通信、数据组织、计算设备、计算流等网络。在数学中,它们在几何和拓扑的某些部分(如纽结理论)中很有用。代数图理论与群论有密切的联系。还有连续的图像,但在大多数情况下,图论的研究属于离散数学的范畴。

运筹学提供了解决商业和其他领域实际问题的技术,如分配资源以实现利润最大化,或安排项目活动,以最大限度地降低风险等问题。运筹学技术包括线性规划和其他领域的优化、排队理论、调度理论、网络理论。运筹学还包括连续时间马尔可夫过程、连续时间鞅(马尔廷加莱斯)、过程优化以及连续和混合控制理论等连续性课题。

虽然拓扑学属于数学领域,它形式化概括了物体的"连续变形"的直观概念,但它产生了许多离散的课题,这在一定程度上可归于对拓扑不变性的关注,而拓扑不变性本身通常采用离散值。参见组合拓扑、拓扑图理论、拓扑组合、计算拓扑、离散拓扑空间、有限拓扑空间。

Text B 树

一个连通图如果没有包含简单的回路,则称为一棵树。早在1857年,英国数学家阿瑟·凯莱就用树计算某些化合物的可能类型。自那时起,树已被用来解决各种科学问题。

树在计算机科学中特别有用。例如,树用来构建高效的在列表中的定位算法。它们也用于构建电话线路网络,使数据存储和传送成本最小。树可用于模拟决策的一系列过程,这使得树的排序算法研究具有非常高的价值。

1. 定义

一棵树是一个无向简单图 G,满足以下任一等价条件。

- G 是连通的,并且没有回路。
- G 没有回路,并且添加任何一条边就会形成一个回路。
- G 是连通的,并且去掉任一条边,它就不是连通图。
- G 是连通的,并且不包含3顶点完全图 K_3 的子图。
- G 中的任意两个顶点都可通过唯一的简单路径连接。

如果 G 有有限个顶点,就说它们是 n 阶的,那么上面的命题也相当于下列任一条件。

- G 是连通的,并具有 $(n-1)$ 条边。
- G 没有简单的回路,并具有 $(n-1)$ 条边。

图3-1所示的图形是树吗?

$G1$ 和 $G2$ 都是树,因为两者都是连通图,没有简单的电路;$G3$ 不是树,因为 e、b、a、d、e 是一个简单的电路。

在该图中,最后,$G4$ 不是树,因为它不是连通的。

2. 性质

(1) n 个顶点的树有 $(n-1)$ 条边。

(2) 一个完全 m-叉树,若含有 i 个内部顶点,则有 $n = m \times i + 1$ 个顶点。

(3) 一个完全 m-叉树:

（ⅰ）若有 n 个顶点,则内部顶点数 $i=(n-1)/m$ 和树叶数 $L=[(m-1)\times n+1]/m$。

（ⅱ）若有 i 个内部顶点,则有 $n=m\times i+1$ 个顶点和 $L=(m-1)\times i+1$ 片树叶。

（ⅲ）若有 L 片树叶,则有 $n=(m\times L-1)/(m-1)$ 个顶点,并且 $i=(L-1)/(m-1)$ 个内部顶点。

3. 树的术语

(1) m-叉树是每一个内部顶点的孩子数不多于 m 个的树。

(2) 二叉树是一个 m-叉树,仅当 $m=2$ 时(每个结点仅有左或右子结点)。

(3) 有序树是一棵树,其中每个内部顶点的孩子顺序是线性排列的。

(4) 平衡树是一个树高为 h 的树,其顶点所在的子树高为 h 或 $h-1$。

(5) 二叉搜索树是一个二叉树,若它的左子树不空,则左子树上所有结点的值均小于它的根结点的值;若它的右子树不空,则右子树上所有结点的值均大于它的根结点的值。

(6) 决策树是有根树,每个顶点代表决定的可能结果,叶子代表可能的解决方案。

(7) 生成树:包含图的所有顶点的树。

(8) 最小生成树:边的权重之和最小的生成树。

4. 树遍历算法

系统访问有序根树的每个顶点的过程称为遍历算法。下面描述 3 种最常用的算法,即前序遍历、中序遍历和后序遍历。

(1) 前序遍历

前序遍历是递归定义的有序根树的顶点列表首先列出其根结点,其次以从左到右出现的顺序列出其第一棵子树,后面跟着其他子树。

若 T 是一个包含根为 r 的有序根树,如果 T 只由 r 组成,那么 r 是 T 的前序遍历。否则,假设 $T_1, T_2, \cdots, T_n$ 是 T 中从左到右的子树,前序遍历从访问 r 开始,继续按前序遍历 T_1,然后按前序遍历 T_2,以此类推,直到 T_n 前序遍历过。

(2) 中序遍历

中序遍历是递归定义的有序根树的顶点列表紧跟在第一个子树的后面,后面再跟着其他子树。它们按从左到右出现的顺序让 T 是有序的根树。

如果 T 只由 r 组成,那么 r 就是 t 的中序遍历。否则,假设 $T_1, T_2, \cdots, T_n$ 是 T 从左到右的子树,按中序顺序遍历从 T_1 开始,访问 r,继续按中序遍历 T_2、T_3,$\cdots$,最后中序遍历 T_n。

(3) 后序遍历

后序遍历是递归定义的有序根树的顶点列表按从左到右出现的顺序列出子树,然后是根目录。

如果 T 只由 r 组成,则让 T 成为根 r 的有序根树,那么 r 就是 T 的后序遍历。否则,假设 $T_1, T_2, \cdots, T_n$ 是从左到右的 r 的子树,后序遍历是从 T_1 按后序遍历开始,然后按后序遍历 $T_2, T_3, \cdots, T_n$,最后访问 r 结束。

Unit 4 软件工程

Text A 软件过程

软件过程是指生产软件产品一系列行为活动集。这些活动集可以包括应用标准编程语言(如 Java 或 C)从设计草图开始的软件开发。然而,越来越多的新的软件开发是扩展和修改已有的系统或配置和集成货架的软件或系统组件。

像所有智力产品和创新过程一样,软件过程是复杂的,依靠人做出决策和判断。因为需要判断和创新,试图软件过程自动化只取得有限的进展。计算机辅助软件工程(CASE)工具可以支持某些开发过程行为。但是,至少在未来的几年中,还没有可能通过更广泛的软件自动化代替软件过程中工程师从事的创意设计。

CASE 工具的有效性有限,是因为软件过程复杂多样、千变万化。由于没有理想的开发过程,许多企业组织都开发了自身特点的软件开发方法。工艺已经发展到利用企业组织中人的能力,以及所开发系统的具体特点。对于某些系统,如关键的系统,需要非常结构化的开发过程。对商业系统,要快速适应需求的变化,灵活、敏捷的开发过程很可能更有效。

尽管有许多种软件过程,但对所有软件进程,都存在一些共同的基本行为活动。

(1) 软件规范:定义软件的功能及其操作上的约束。
(2) 软件设计和实现:将满足规范要求的软件生产出来。
(3) 软件验证:必须验证软件是否满足客户的需要。
(4) 软件进化:软件必须不断发展,以满足不断变化的客户需求。

没有"理想的"软件过程,许多组织都存在用于改进软件过程的框架。这些过程可能包括过时的技术或可能在工业软件工程最佳实践中不占优势。事实上,很多企业组织仍然在他们的软件开发中没有充分利用好软件工程方法。

软件过程可以通过过程标准化减少不同企业组织的软件过程的多样性,这将提高沟通性,减少培训时间,使流程自动化,也更经济。标准化也是引进新的软件工程方法和技术,进行良好软件工程实践的重要的第一步。

1. 软件过程模型

软件过程模型是软件过程的抽象表示。每个过程模型代表从一个特定的角度的过程,从而提供了有关该过程某部分的信息。本节介绍一些很通用的流程模型(有时被称为过程范例),并从架构体系的角度呈现。也就是说,我们看到的是该过程的框架,而不是具体活动的细节。

这些通用模型不是软件过程的定义描述。相反,它们是可用于解释不同的软件开发过程中的抽象。你可以把它们看作过程框架,可扩展适用于更具体的软件工程过程。

(1) 瀑布模型,这种模型定义说明、开发、验证和进化等基本过程的行为活动,并将它们作为独立的过程,如需求说明、软件设计、实现、测试等。

(2) 进化开发模型,这种模型将说明、开发和验证的活动交织在一起。初始系统是从抽象的说明快速开发的,然后根据客户(需求)输入不断精炼,产生满足客户需求的系统。

（3）基于组件的软件工程模型，这种模型用于基于大量可重用组件的软件开发。该系统的开发过程着重于将这些组件集成到一个系统，而不是从头开始开发它们。

2. 瀑布模型

第一个发布的软件开发过程模型是从更一般化的系统工程过程推导出的（劳斯莱斯，1970年），如图4-1所示。因为一个阶段与另一个阶段嵌套在一起，所以这种模式被称为瀑布模型或软件生命周期。模型的主要阶段对应软件开发的基本行为活动。

ⅰ．需求分析和定义　通过与系统用户协商建立系统的服务、约束和目标，然后详细定义它们，并作为系统的规范说明。

ⅱ．系统和软件设计　系统设计过程中将需求分配到硬件或软件系统上。它建立了系统的整体架构。软件设计包括识别和描述基本软件系统的功能抽象和它们之间的关系。

ⅲ．实现和单元测试　这个阶段，软件设计内容由一组程序或程序单元实现。单元测试涉及验证每个单元是否符合其规范。

ⅳ．集成和系统测试　各个程序单元或程序集成在一起，构成一个完整的系统进行测试，以确保软件的需求已得到满足。经过测试，该软件系统就交付给客户。

ⅴ．操作和维护　通常（尽管不是必需的），这是最长的生命周期阶段。系统安装并投入实际使用。维护包括改正在生命周期的早期阶段没有被发现的错误，改进系统单元的实现，在新需求提出后增加系统的服务。

原则上，每个阶段的结果是经批准的一个或多个文档（签名关闭）。后续的阶段不应该启动，直到前一个阶段已经完成。在实践中，这些阶段相互重叠，彼此间反馈信息。在设计阶段，发现需求中的问题；在编码中，设计存在问题被发现等。软件过程不是一个简单的线性模型，而是涉及开发活动的反复迭代序列。

由于文件制定和批准需要成本，涉及明显返工的反复迭代工作代价大，因此经过少量反复迭代工作后，通常冻结部分开发工作（如说明书），然后才继续后面的开发阶段。留下的问题接着细化、忽略或编程规避。这种过早的需求冻结可能意味着该系统将不是用户想要的。实现过程的技巧规避也可能导致严重的结构化系统设计问题。

在最后的生命周期阶段（操作和维护），软件被投入使用。原始的软件需求存在的错误和遗漏将被发现。程序和设计错误将涌现，新的功能需求也将确立。因此，该系统必须进化，以保持实用性。进行这些更改（软件维护）可能涉及重复前面的过程各个阶段。

瀑布模型的优点是文档在各个阶段产生，它与其他工程过程模型相吻合。其主要问题是在项目中划分出的不同的阶段，其灵活性不够。必须在软件过程的早期阶段给出很好的保证，使得其难以满足不断变化的顾客需求。

因此，瀑布模型适应于在系统开发中需求易于理解和不太变化的项目。

然而，瀑布模型反映了在其他工程项目中使用的过程模型类的特点。因此，基于这种方法的软件过程模型也用于软件开发，特别是当软件项目是一个更大的系统工程项目的一部分时。

Text B UML 简介

统一建模语言(UML)是编写软件蓝图的标准语言。UML 可用于软件密集型系统产品可视化描述、规范、构建和文档化。

UML 适合于系统建模,从企业信息系统到基于 Web 的分布式应用程序,甚至硬实时嵌入式系统。这是一个非常传神的语言,涉及开发,然后部署这样的系统所需的所有方面。虽然 UML 功能强大,但是它不是很难理解和使用。要学习使用 UML,一个有效的出发点就是形成该语言的概念模型,这就要求学习三大要素:UML 的基本结构块、支配这些结构块如何放置在一起的规则以及应用于整个语言的一些公用机制。

由于 UML 仅是一门语言,因此它仅是一种软件开发方法的一部分。UML 是过程独立的,虽然对用例驱动、以架构为中心、迭代和增量过程要最佳化。

1. UML 概述

UML 是一门语言,适用于可视化、规范化、结构化、文档化。

2. UML 是一门语言

语言包括一个词汇表以及组合词汇进行交流的规则。建模语言是一门语言,它的词汇表和规则集中在一个系统的概念和物理表示上。建模语言(如 UML)是适合描述软件蓝图的标准语言。

建模需要了解一个系统。没有用哪一种模型描述就足够了。相反,除了细小的系统,你经常需要彼此相关的多个模型深入了解(系统特性)。对于软件密集型系统,这需要一门语言从不同的角度描述在整个软件开发生命周期的系统的体系结构。

如 UML 的词汇表和规则,告诉你如何创建和读取结构良好的模型,但它们不告诉你,你应该建立什么样的模型,什么时候应该创建它们。那是软件开发过程中的功能。一个定义明确的过程将指导你生产什么制品,有什么活动和使用什么样的工人创建它们并对其进行管理,以及如何使用这些制品作为一个整体评估和控制项目(进展)。

3. UML 是可视化语言

对于很多程序员,思考代码实现到具体编码之间的距离接近零。你一边想,一边编码。事实上,有些事情最好直接由代码描述。文字是写表达式和算法最奇妙的简洁和直接的方式。

在这些情景下,程序员还是在做一些建模工作,尽管比较弱智。他或她甚至可能在白色板或餐巾纸上勾勒出一些想法。但是,这样做存在几个问题:首先,与别人交流这些概念模型时容易出错,除非涉及的人都讲同一种语言。一般情况下,项目组和企业组织都开发他们自己的语言,如果你是一个局外人或你新加入项目组,就很难理解发生了什么事情。其次,有关一个软件系统的东西,还存在一些你无法理解的事情,除非你建立超越文本编程语言模型。例如,通过分析层次结构中的所有类的代码,可以推断出类层次结构的含义,但不能直接领会。同样,通过研究系统的代码,可以推断出基于 Web 系统的对象的物理分布和可能的迁移活动,但不能直接领会。其三,如果开发者砍掉那些代码,对应在他或她的脑袋里的模型又从来没有写下来,这些信息将永远失去,或者充其量只是部分可重建,由开发者继续实现一次。

采用 UML 建模解决了第三个问题：一个简明的模型有利于沟通。

有些事情最好以文本为蓝本；其他的最好图形化建模。事实上，在所有的有意义的系统中，有超越用编程语言表示的结构。UML 就是这样一种图形化的语言，可解决上述的第二个问题。

UML 不仅仅是一堆图形符号集。相反，在 UML 每个符号的背后，具有一个良好定义的语义。在这种方式下，一个开发人员可以编写 UML 模型，另一名开发人员，甚至其他工具，可以无歧异地解释该模型。这就解决了上述的第一个问题。

4. UML 是实现规范化的语言

本文中，规范化的意思是建立精确的、明确的和完整的模型。特别是，UML 解决了在开发和部署软件密集型系统需要进行的所有重要的分析、设计和实现决策的规范化问题。

5. UML 是一种结构化语言

UML 不是一种可视化的编程语言，但它的模型可以直接关联到各种编程语言。这意味着，它可以从 UML 模型映射到编程语言，如 Java、C++ 或 Visual Basic，甚至映射到关系数据库中的表或面向对象的数据库永久存储对象。在 UML 中，事情最好用图表表示，而在编程语言中，事情最好用文本表达。

这个映射允许从 UML 模型到编程语言的正向工程中生成代码。反过来也是可能的：你可以从代码实现到 UML 建立模型。逆向工程不是魔术，除非你在执行中对信息进行编码，否则在模型转换代码时，信息就丢失了。逆向工程需要与人机交互作用工具的支持。结合正向工程代码生成和逆向工程这两条途径，就可以产生双向工程，这意味着既能在图形视图下工作，也可以在文本视图下工作，只要用工具保持二者的一致性即可。

除了这种直接的映射，UML 还有足够的和明确的表现力，允许直接描述模型、系统仿真和运行系统检测。

6. UML 是一种文档化语言

一个良好的软件企业除了提供可执行的代码外，还可以生产各种各样的智力产品。这些产品包括（但不限于）需求、架构、设计、源代码、项目计划、试验、原型、发布。

根据不同的开发习惯，这些产品为越来越多的人所需要，虽不是正式的。这样的产品不仅是一个项目的可交付成果，也是软件系统开发和部署后进行控制、检测和交流的关键。

UML 强调系统的体系结构文档及其所有的细节。UML 是满足测试需求的描述语言。最后，UML 也是满足项目规划和发布管理等活动所需的建模语言。

Unit 5 数　据　库

Text A　MySQL 简介

MySQL 数据库系统使用客户机/服务器结构，主要安装在服务器上。服务器是实际操纵数据库的程序。客户端程序不直接这样做。相反，它们使用结构化查询语言（SQL）语句传达你的意图到服务器。客户端程序安装在本地的你要访问 MySQL 的机器上，但服务器可以在任何地方进行安装，只要客户端可以连接到它。MySQL 是一种固有的联网数据库系统，以便客户端可以与服务器进行通信，服务器可以运行在本地计算机或其他地方的机器上，也可能在地球的另一边。客户机可以因许多不同的功能需要而编写，但每一个连接到服务器的交互过程，包括发送 SQL 语句到服务器执行数据库操作，并从它接收该语句的结果。

这样的一个 MySQL 程序客户端包含在 MySQL 发布包中。当交互使用时，MySQL 提示你输入语句，将其发送到 MySQL 服务器执行，然后显示结果。这种功能非常有用，使 MySQL 具有自己的特色，但它也是一个宝贵的工具，通过 MySQL 编程活动帮助你。这往往便于从一个脚本程序内快速查看访问表结构，在程序中使用它之前尝试一条语句，确保它产生一种正确的输出，等等。MySQL 非常适合做这些工作。MySQL 也可以非交互使用，例如从文件中或从其他程序语句读取。这使你可以将 MySQL 使用在脚本中或时钟进程中或与其他应用程序相结合。

通过本章的介绍，就可以更有效地使用 MySQL 的功能。
- 启动和停止 MySQL。
- 指定连接参数，并使用选项文件。
- 设置 PATH 变量，以便你的命令解释器能找到 MySQL（和其他 MySQL 程序）。
- 交互发出 SQL 语句和使用批处理文件。
- 取消和编辑语言命令。
- 控制 MySQL 的输出格式。

要使用所示的例子，需要一个 MySQL 用户账户和一个工作数据库。本章的前两部分描述了如何使用 MySQL 设置这些。为达到演示目的，假设你使用的 MySQL 例子如下。
- MySQL 服务器在本地主机上运行。
- 你的 MySQL 用户名和密码分别是 cbuser 和 cbpass。
- 你的数据库名为 cookbook。

对你自己的实验，可以违背这些假设。您的服务器不必本地运行，并且无须使用在本章使用的用户名、密码或数据库名称。自然，如果你的系统上使用的是不同的默认值，就需要修改例子的对应内容。

即使不使用 cookbook 作为你的数据库名称，建议你创建一个专用数据库专门尝试这个例子，而不是在你目前用作其他用途的数据库进行尝试。否则，现有表的名称可能与范例中的名称冲突，你必须做出修改，当你使用一个单独的数据库时，就不必要了。

如果你有最喜欢用于发出查询的另一个客户端程序，本章中的一些概念可能不适用。例如，你可能更喜欢图形化 MySQL 查询浏览器的程序，它提供了一个点击鼠标界面的 MySQL 数据库。在这种情况下，一些原则将是不同的，如你终止 SQL 语句的方式。在 MySQL 中，终止语句用分号（;）字符，而在 MySQL 查询浏览器中有"终止语句"的执行按钮。另一种流行的界面是 phpMyAdmin，它使你能够通过 Web 浏览器访问 MySQL。

Text B　执行事务

因为 MySQL 服务器是多线程，所以它可以在同一时间处理多个客户端请求。为了应对客户端之间的争夺,服务器对两个客户端不能同时修改同一数据进行任何必要的锁定。然而,当服务器执行 SQL 语句时,它很可能是连续地从给定的客户机接收语句,并与来自其他客户端的语句交织在一起。如果客户端发送的多条语句彼此依赖,事实上,其他客户端执行这些语句更新的数据表就会陷入困境。如果多条语句的操作不能运行到完成,语句失败也可能会有问题。假设你有一个包含有关航空公司航班安排信息的飞行表,你想从那些可用的人员中选择一个飞行员,更新航班 578 记录。你可以使用如下 3 条语句。

```
SELECT @p_val := pilot_id FROM pilot WHERE available = 'yes' LIMIT 1;
UPDATE pilot SET available = 'no' WHERE pilot_id = @p_val;
UPDATE flight SET pilot_id = @p_val WHERE flight_id = 578;
```

第一条语句选择一个可用飞行员,第二条语句标记其为不可用,第三条语句将该飞行员分配到该航班。这在实践中足够直截了当,但在原理上,这个过程存在一系列问题:

如果两个客户要安排飞行员,可能他们都将运行初始的 SELECT 查询语句,在设置飞行员的状态为不可用之前,他们有机会检索到相同的飞行员 ID 号。如果出现这种情况,同一飞行员就会被立刻安排到两个航班上。

这三条语句必须作为一个整体成功执行。例如,如果 SELECT 语句和第一条 UPDATE 语句成功运行,但第二条 UPDATE 语句失败,则飞行员的状态设置为不可用,而又没有将飞行员分配到一个航班。该数据库将处于不一致的状态。

为了防止在这类情况下的并发性和完整性问题,事务(处理)概念是有帮助的。一个事务是一组语句,并确保以下特性:

- 在事务过程中没有其他客户端可以更新在事务处理中使用的数据。例如,当你预订航班飞行员时,其他客户不能修改飞行员或飞行记录。在你执行操作时,通过防止其他客户干扰,事务处理解决了 MySQL 服务器多用户而产生的并发问题。事务处理串行化多条语句操作访问共享资源。
- 事务中的语句组成为一个单位提交起作用(生效),当且仅当他们全部成功时。如果发生错误,错误前面发生的任何动作被回滚,留下相关的表不受影响,就好像根本没有任何语句发出一样。这样可以保证数据库不会变得不一致。例如,如果一个更新航班表语句失败,回滚导致改变飞行员表操作撤销,而使驾驶员仍然可用。回滚操作使得你不必清楚如何撤销部分已完成的操作。

Unit 6 嵌入式系统

Text A 什么是嵌入式系统

过去几十年,一个令人惊讶的发展是计算机在人类事物中崛起占主导位置。今天,在我们的家庭和办公室,计算机台数已比生活和工作其中的人数更多。然而,许多计算机都没有被它们的用户看出来。本单元将解释什么是嵌入式系统,在哪里可以找到它,还将介绍嵌入式编程的知识,解释为什么本书选择使用 C 和 C++ 语言,描述范例中所使用到的硬件。

1. 关于嵌入式系统

嵌入式系统是计算机硬件和软件，以及可能的附加机械或其他部件设计成执行特定功能的组合。一个很好的例子是微波炉。几乎每家每户都有一个微波炉，数以千万计的人每天都在使用它，但很少有人意识到，一个处理器和软件都参与了他们的午餐或晚餐的准备工作。

与家用的个人计算机对比，它也是由计算机硬件和软件，以及机械部件（如磁盘驱动器）组成。然而，个人计算机不设计成执行特定的功能。相反，它能够做许多不同的事情。许多人使用通用计算机这个词进行明确的区分。装货时，通用计算机处于一个空白状态；制造商不知道客户将用它做什么。一个客户可能会用它作网络文件服务器，另一个客户可能用它玩游戏，第三个客户可能用它编写下一部宏伟的美国小说。

通常，一个嵌入式系统是一些较大系统的一个部件。例如，现代汽车和卡车包含许多嵌入式系统。一个嵌入式系统控制防抱死制动系统，另一个嵌入式系统监视并控制车辆的排放量，第三个嵌入式系统在仪表板显示信息。在某些情况下，这些嵌入式系统通过某种通信网络连接起来，但这当然不是必需的。

这可能令你迷惑，必须指出，一个通用计算机本身就是由大量的嵌入式系统组成的。例如，我的计算机由键盘、鼠标、视频卡、调制解调器、硬盘驱动器、软盘驱动器和声卡组成，它们每一个都是嵌入式系统。这些设备都包括处理器和软件，并且设计成执行特定的功能。例如，调制解调器被设计成通过模拟电话线发送和接收数字数据。就是这样而已。和所有其他设备一样，可以概括成一句话。

如果一个嵌入式系统设计得很好，所用设备的用户可能完全感觉不到处理器和软件的存在，如微波炉、VCR 或闹钟。在某些情况下，甚至有可能建造一个不包含处理器和软件的等效设备。这可以通过定制的集成电路取代执行相同功能的硬件组合完成。然而，当设计成这种方式的硬件时，会丢失很大的灵活性。比起重新设计一块定制的硬件，改变几行软件代码很容易、更便宜。

2. 历史与未来

直到 1971 年，第一个这样的系统才出现。那年，英特尔推出了全球第一个微处理器。芯片 4004 被设计成在日本 Busicom 公司生产的商务计算器使用。1969 年，Busicom 要求英特尔设计出一套定制的集成电路，使用在它们的新计算器模型上。4004 芯片是英特尔公司对需求的回应。而不是为每个计算器设计定制硬件，英特尔建议，一个通用电路可以在整个计算器行业中使用。这种通用处理器设计成读取和执行一组指令的软件，存储在外部的存储芯片中。英特尔的想法是，该软件可以给每台计算器其独特的功能集。

这个微处理器一夜成名，它的应用在随后的十年稳步增长。最早的嵌入式应用包括无人驾驶的空间探测器、计算机操作的交通信号灯，以及飞机的飞行控制系统。20 世纪 80 年代，嵌入式系统悄悄踏着微计算机时代的巨浪将微处理器带到我们个人和职业生活的每一个角落。许多电子设备用在厨房（面包机、食品加工器和微波炉）、起居室（电视机、音响和遥控器）和工作场所（传真机、寻呼机、激光打印机、收银机和信用卡读卡器）的嵌入式系统中。

嵌入式系统的数量将继续迅速增加，似乎势不可挡。已经有前景光明、有巨大市场潜力的新嵌入式设备可通过中央计算机控制灯的开关和恒温器,当孩子或小个子成年人在现场时不进行充气的智能气囊,手掌大小的电子设备和个人数字助理机(PDAs),数码相机和仪表板导航系统。显然,将来相当长的一段时间,迫切需要个人拥有设计下一代嵌入式系统的技能和期望。

3. 实时系统

这一嵌入式系统的子类是值得介绍的。如通常定义的实时系统是具有时间约束的计算机系统。换言之,一个实时系统中的某些计算或决策能力是与时间紧密设定的。这些重要的计算据说有完成的截止时间要求。而且,对于所有的实际应用,错过截止时间与一个错误的答案一样糟糕。

如果一个截止时间错过了,会发生什么问题,这是一个关键问题。例如,如果实时系统是飞机的飞行控制系统的一部分,单一错过截止时间问题,它有可能给乘客和机组人员的生命带来威胁。然而,如果所涉及的系统是卫星通信,损害可能被限制在单个损坏的数据包。越严重的后果,就越有可能说截止时间是"硬"的,因此,该系统是硬实时系统。在时间连续的另一端实时系统,就认为具有"软"的截止时间。

一个实时系统的设计者必须在他的领域更加勤奋工作。他必须保证在所有可能的条件下,软件和硬件可靠地运行,并且对人的生命依赖于系统的正确执行程度,必须通过工程计算和文字描述加以论证来确保。

Text B 认识硬件

嵌入式软件工程师在自己的职业生涯中会有机会与许多不同的硬件一起工作。在本章中,我会教你一个简单过程,我是如何熟悉新的电路板的。在这个过程中,我会指导你通过创建一个头文件描述主板最重要的特点和初始化硬件到已知状态的软件。

1. 了解大概

编写嵌入式系统软件之前,必须先熟悉在其上运行的硬件。首先,你只需要了解系统的一般操作。不需要了解硬件的每一个细节;那些知识不会立刻就需要,会随着时间的推移而熟悉。

每当收到一个新电路板,你应该花一些时间阅读它提供的所有文档。如果电路板是一个现成的产品,它可能伴随有充分考虑了软件开发人员需求的"用户手册"或"程序员手册"。但是,如果这个电路板是为你的项目定制设计的,该文档可能会更神秘或主要为硬件设计师参考而写。无论哪种方式,这(阅读文档资料)是你开始工作的最好地方。

当正在阅读文档时,请将电路板放在一边。这将有助于你专注大局。读完文档后你会有充足的时间更仔细地检查实际的电路板。拿起电路板前,应该能够回答有关它的两个基本问题。

- 电路板的总体功能是什么?
- 数据如何通过它(交换)?

例如,假设你是调制解调器设计团队中的一员。你是软件开发人员,刚刚从硬件设计师那里收到早期原型板。因为你已经熟悉了调制解调器,对你而言,它的总体功能和通过的数据流应该是相当清楚的。板子的目的是通过模拟电话线发送和接收数字数据。硬件从一组电气连接中读取数字数据和写入模拟形式的数据到连接的电话线路上。数据流还存在相反的方向，从电话线插口读入模拟数据和输出数字数据。

虽然大多数系统的功能是相当明显的，但是数据流可能不是。我经常发现一个数据流方框图对实现快速理解是有帮助的。如果你很幸运，提供给你的硬件文档将包含你所需要的框图的一个超集。然而，你可能仍然发现创建自己的数据流图还是有用的。这样，你可以忽略系统中与基本数据流程不相关的硬件组件。

对于 Arcom 公司生产的板子，该硬件设计时没有考虑到一个特定的应用。因此，对于本章的剩余部分，我们将不得不想象，它确实有一个目的。我们应当假定该电路板设计用作打印机共享设备。打印机共享设备允许两台计算机共享一台打印机。

该装置的用户连接计算机到每个串口和一台打印机连到并行口。然后两台计算机都可以将文档发送到打印机，虽然某一时刻只能有一台计算机可以运行。

为了说明数据是如何通过打印机共享设备流动的，可以画框图，如图 6-1 所示（仅画出了所涉及本应用的 Arcom 板子上的那些硬件装置）。通过查看框图，你应该能够将通过该系统的数据流快速可视化。要打印的数据从任一串口接收，保存在 RAM 中，直到打印机准备接收更多的数据，并通过并行端口传送到打印机。实现这一切功能的软件都存储在 ROM 中。

一旦创建了一个框图，不要弄皱并把它扔掉，应该把它放好，在整个项目中，你都可以参考它。建议创建一个项目笔记本，或将其粘在第一页上。随着你继续使用这块硬件板，在笔记本上写下你所了解的一切。你可能还希望保存有关软件设计和实现的记录。项目笔记本是有价值的，不仅在你正在开发软件时，而且在该项目已完成后。几个月或几年后，当你需要更改你的软件，或处理类似的硬件时，你将感激你付出额外的努力，记录到笔记本上。

2. 检查外观

设想自己处在处理器的位置一会儿，往往是有益的。毕竟，处理器最终只会做你的软件指示它做的。想象一下处理器是什么样的：处理器的世界会是什么样子？如果从这个角度想，你很快意识到一件事是，处理器有很多伙伴。这些都是电路板上的其他硬件部分，与该处理器可以直接通信。在本节中，你将学会识别它们的姓名和地址。

首先要注意的事情是，有两种基本类型：存储器和外设。显然，存储器用于数据和代码存储和检索。但你可能想知道外设是什么。这些专门的硬件设备，要么通过（I/O）与外面的世界相互作用，或者执行特定的硬件功能。例如，在嵌入式系统中，两个最常见的外设是串行端口和定时器。前者是一个 I/O 设备，后者基本上是一个计数器。

Intel 的 80x86 的成员和其他一些处理器家族有两个不同的地址空间，通过它们可以在存储器和外设间进行通信。第一类地址空间称为存储器空间，并主要用于存储器装置；第二类只保留给外围设备，称为 I/O 空间。然而，外围设备也可以位于存储器空间内，由硬件设计者决定。当这种情况发生时，我们说这些外设是内存映射。

从处理器的角度看，内存映射外设的外观和行为非常像存储设备。然而，外围设备的功能很明显与一个存储器的作用不同，并不是简单的存储提供给它的数据，外设可能代替数据解释为命令，或以某种方式处理数据。如果外设位于存储空间内，我们就说系统内存映射 I/O。

嵌入式硬件的设计人员往往只喜欢使用内存映射 I/O,因为对硬件和软件开发者都有优势。它对硬件开发者有吸引力,因为它可能消除 I/O 空间,及其相关联的一些导线。这不仅可显著降低电路板的生产成本,也可降低硬件设计的复杂度。对程序员内存映射的外设也是有益的,能够使用指针、数据结构和联合体,更容易、更有效地与外设进行交互。

Unit 7　计算机网络

Text A　因特网协议族

因特网(Internet)协议族是一组用于互联网和其他类似网络的通信协议。它通常也称为 TCP/IP,取自两个最重要的协议命名:传输控制协议(TCP)和网际协议(IP),这是在该标准中定义的前两个网络协议。现代 IP 网络的代表始于 20 世纪 60 年代和 70 年代,20 世纪 80 年代演变出互联网和局域网,20 世纪 90 年代初出现万维网网站。

互联网协议族像许多协议族,由多层次组成。每一层解决涉及数据的一组传输问题。特别是,每层定义层内协议的操作范围。

通常,某层的组件为上层协议提供定义明确的服务,并且服务可来自其低层的服务。上层逻辑上贴近用户和处理更抽象的数据,依靠低层协议翻译,将数据转换成最终适宜物理传输的形式。

TCP/IP 模型包括 4 层,从低到高依次是链路层、网络层、传输层以及应用层。

1. 发展史

互联网协议族源于 1970 年初美国国防高级研究计划局(DARPA)进行的研究和开发。1973 年春天,已有的 ARPANET 网络控制程序(NCP)协议的开发者温顿·瑟夫加入了卡恩主持的开放式架构互连模型工作组,以设计下一代 ARPANET 协议为目标。

截至 1973 年夏天,卡恩和瑟夫曾利用公共互联网络协议重新塑造了一个基本的网络,其隐藏网络协议之间的差异性,而其可靠性不由网络提供,如在 ARPANET 中,是由主机负责的。

1975 年,在斯坦福大学和伦敦大学学院(UCL)两网之间进行了 TCP/IP 通信测试。1977 年 11 月,在美国、英国和挪威站点之间进行了 3 个 TCP/IP 网络互联的试验。1978—1983 年,多个研究中心开发了其他一些 TCP/IP 原型。1983 年 1 月 1 日,当新的协议正式开发完成后就永久激活,ARPANET 迁移到 TCP/IP。

1982 年 3 月,美国国防部宣布 TCP/IP 作为所有军用计算机网络的标准。1985 年,互联网架构委员会为计算机行业举办了为期三天有关 TCP/IP 的研讨会,250 个供应商代表出席,促进协议推广,引领其日益增长的商业应用。

2. 各层互联网协议

TCP/IP 协议族对提供的协议和服务进行了抽象封装。这样的封装通常使协议族对应每一功能层进行划分。一般地,应用程序(该模型的最高级别)使用一组协议向下一层发送其数据,在下一层级进一步封装。

这可以通过一例网络情景进行说明,其中两个因特网计算机主机跨越本地网的边界,即互联网络网关(路由器)进行通信,如图7-1所示。

协议和方法的功能组在应用层、传输层、因特网层以及链路层。这个模型不打算成为一个严格的参考模型,而使用必须适合成为标准的新协议。封装的应用层数据向下通过的协议栈如图7-2所示。

表7-2提供了在各分层对应各自协议族的一些例子。

3. 实现

现今使用的大多数计算机操作系统,包括所有的普通消费定位的系统,都有一个TCP/IP实现。最少可接受的实现包括实现(从最必需的到不太重要的)IP、ARP、ICMP、UDP、TCP和某个IGMP。

程序员可接受的大多数IP实现,都使用抽象套接字(SOCKET)(与其他协议也可以使用)和合适的API用于大部分操作。这个接口称为BSD套接字,并最初用于C语言。

Text B 云计算

1. 什么是云计算

云计算是一种资源交付和应用模型;这意味着通过网络获取资源(如硬件、软件)。提供资源的网络被称为"云"。"云"硬件资源似乎可伸缩无限,并且无论何时都可使用。

云计算既指通过互联网提供服务的应用程序,也包括数据中心提供这些服务的硬件和系统软件。服务本身一直称为软件即服务(SaaS),所以我们用这个词。数据中心硬件和软件,我们称之为云。

2. 新的应用机遇

(1)移动交互式应用程序。Tim O'Reilly认为,"未来属于能实时做出响应的信息服务,无论是它们的用户提供,或是非人传感器提供。"这样的服务将吸引到云,不仅是因为它们必须具有高可用性,而且还因为这些服务通常依赖于海量的数据集,适宜在大型数据中心最方便地承载。这是特别的服务情景,组合两个或多个数据源或其他服务,如混聚应用(mash-ups)。虽然不是所有的移动设备都能够有100%的时间享受连接云计算,但在特定的应用领域已经成功解决断开操作存在的挑战,所以我们不认为这是一个出现移动应用的明显障碍。

(2)并行批处理。到目前为止我们已经关注使用云计算的交互SaaS,云计算提供了批量处理和分析TB级数据的独特机会,并可能花费几个小时才能完成这样的工作。如果应用数据有足够的并行性,用户就可以利用云计算新的"成本相关性"优势:在很短的时间内使用数百台计算机的成本与使用几台计算机很长一段时间相当。例如,《华盛顿邮报》高级工程师彼得·哈金斯利用200个EC2实例(1407服务器小时)将17 481页的希拉里·克林顿旅行证件转换为更友好的形式,以便在WWW发布后的9小时内在WWW上使用。像Google的map reduce及其开源对应的Hadoop这样的编程抽象允许程序员表达这些任务,同时隐藏了在数百台云计算服务器上编排并行执行的操作复杂性。事实上,"云时代"追求在这个领域的商业机会。同样,使用Gray的洞察力,成本/效益分析必须权衡大数据集移动到云,能够加速数据分析,从中获得潜在收益。当回到经济模型后,我们推测,亚马逊提供大型公共数据集免费入驻的部分动机可能会减轻这种成本分析因素,从而吸引用户购买数据处理所需的云计算周期。

（3）分析业的兴起。批量处理的计算密集型特例是业务分析。大型数据库行业最初以事务处理为主,这种需求正在逐渐趋于平稳。越来越多的计算资源,现在用来理解客户、供应链、购买习惯、排名等。因此,虽然网上交易量继续缓慢增长,但是决策支持正在迅速增长,资源的移动倾向是从交易到商业分析的数据库处理。

（4）计算密集型的桌面应用程序扩展。数学的最新版本软件包 MATLAB 和 Mathematica 都能够使用云计算执行费时的计算。其他桌面应用程序同样可能从无缝扩展到云中受益。同样,一个合理的测试结果要对比在云中的计算加上移动数据进出云的成本,减去使用云计算节省时间的成本。符号数学涉及单位计算量巨大,使其成为值得研究的一个领域。一个有趣的替代模型可能是数据保留在云中,并依靠具有足够的带宽,让其适当可视化,返回给人类用户图形化用户界面。脱机图像渲染或 3D 动画可能是一个类似的例子：给出一个三维场景中的对象与所述照明光源特性的紧凑描绘,渲染图像是计算字节比率很高的令人为难的并行计算任务。

（5）"地面上"的应用。某些应用不能适应云的弹性和并行性,可能受挫于数据移动成本、受制于进入和离开云必需的延迟。例如,虽然进行长期财务决策相关的分析适用于云计算,但买卖股票、需要微秒精度则不适用。直到海量数据传输的成本(以及可能的延迟)减少(见第 7 节),这些应用程序可能是明显不适合云计算的代表。

i. 云计算经济学

长期看,决定是否在云托管服务是有道理的,我们认为,使用云计算颗粒度经济模型做出权衡决策更加流畅,尤其是通过云提供的弹性服务转移风险。

同时,硬件资源的成本不断下降,它们以变化的速率发生。例如,计算和存储成本比 WAN 成本下降更快。云计算可以跟踪这些变化,将它们更有效地回馈到更多的客户,将比建造他们自己的数据中心花费更符合实际使用资源的支出。

在决定是否将现有服务转向云计算时,必须仔细考虑预期的平均资源使用和峰值资源使用情况,特别是在应用具有对资源需求高可变性情况下；必须考虑购买设备的实际使用情况；以及在不同云环境下的运行操作成本。

ii. 关于明天的云

长期以来,梦想中的计算作为一种实用工具的愿景终于出现了。通过互联网这种实用工具的可塑性,直接向客户提供与业务需求相匹配的服务,因为工作负载的增长(和收缩)速度可能远远快于 20 年前。过去将业务发展到数百万客户需要数年时间,现在可能几个月内就能实现。

有人质疑公司是否习惯了高利润的企业,如搜索引擎和传统的套装软件广告收入,可以在云计算竞争。首先,这个问题假定云计算是一种基于其低成本小利润的业务。鉴于中型数据中心的典型利用率,规模经济的潜在因子 5:7,并在云计算数据中心的地点选择上进一步节省,提供给云用户显然成本低,云服务提供商仍然可能是非常有利可图的。其次,这些公司可能在他们的主线业务的基础设施中已经拥有数据中心、网络和软件。因此,云计算模式代表再付出一点额外的费用获得更多收入的机会。

Unit 8 数 据 结 构

Text A　Java 语言的数据结构和算法

一旦学会了编程,就会遇到现实世界中的问题,解决这些问题需要的不仅是一门编程语言。Java 语言的数据结构与算法就是一个逐渐沉浸其中的最实用的方式,它可使数据做你想让它做的事情。

1. 数据结构概貌

看数据结构的另一种角度是专注于它的长处和短处。本节将提供一个概貌,如表 8-1 所示,我们将讨论主要的数据存储结构。

2. 算法概貌

我们讨论的很多算法都直接适用于特定的数据结构。对于大多数的数据结构,我们需要知道

(1) 插入新的数据项。

(2) 搜索指定的项。

(3) 删除指定的项。

可能还需要知道如何通过迭代遍历数据结构中的所有项目,因此访问每一项,以显示它或在其上执行一些其他操作。其中一个重要的算法类称为排序。对数据进行排序有很多种方法,递归的概念在设计某些算法时很重要。

3. 面向对象编程

因为过程语言如 C、Pascal 和 BASIC 不能处理大型和复杂的程序,因此要发明面向对象编程语言。为什么这样?

这些问题与程序的整体组织有关。过程编程通过将代码划分为一个个函数(一些语言称为过程或子程序)组成。一组函数可以形成较大的单位称为模块或文件。基于函数组成的代码存在的一个问题是以牺牲数据为代价来关注函数。对数据,没有很多的选择。简单地说,数据对本地的一个特定的函数可访问,或者它对所有函数全局可访问。没有办法(至少没有灵活的方式)指定某些函数可以访问一个变量,而其他函数不能。当几个函数需要访问相同的数据时将引起问题。要提供给一个以上的函数访问,这样的变量必须是全局性的,但全局数据可能被程序中的任何函数无意访问到。这导致频繁的编程错误。所需要做的是微调数据的访问方式,允许变量对需要访问它的函数有效,但对其他函数隐藏起来。

4. 简而言之

对象概念的想法产生于编程社区,作为解决过程语言问题的方案。

(1) 对象

对象是新的实体。对象可以同时解决几个问题。不仅一个编程对象更精确地与真实世界中的对象相对应,而且它也解决了过程模型中的全局数据所产生的问题。

(2) 类

你可能认为一个"对象"的想法足以引起编程革命,但还有更多的新思想。早期,人

们认识到,你可能想使用相同类型的多个对象。也许你正在为整个公寓编写一炉温控制程序,例如,你需要在你的程序中添加几十个温控对象。因此,类的想法诞生了。

Text B　两类重要数据结构

这一部分将简要介绍两个重要的数据结构:链表和哈希表,这是有效搜索或存储数据的结构。

1. 链表

我们看到,作为数据存储结构,数组有一定的缺点。对无序排列的数组,搜索是缓慢的,而在有序数组中,插入是缓慢的。在这两种情况下进行数组删除操作速度很慢。另外,数组创建之后,大小不能变化。在这一部分,我们将看看能解决一些问题的数据存储结构:链表。

链表可能是数组之后的第二个最常用的通用存储结构。链表是一种多功能的结构,适用于多种通用数据库。它也可以取代数组,作为其他存储结构,如堆栈和队列的基础。事实上,在很多情况下可以使用一个链表代替数组(除非你需要使用一个索引频繁且随机存取个别元素)。链表并不能解决所有的数据存储问题,但它们提供令人惊讶的功能和概念,比其他一些流行的结构(如树结构)简单。因为我们专注发现它们的长处和短处。在这一节中我们参考两个非常有用的数据结构。

2. 链接

在一个链表中,各数据项嵌入在一个链接。链接是一个类似链条的对象。因为链接与列表链接有很多类似性,所以使用一个单独的类从链表本身区分出来是有道理的。每个链接对象都包含一个引用(通常称为下一个next),通常指向列表中的下一链接对象。包含列表的字段本身参考第一链路。下面是一个链接类定义的一部分,它包含一些数据,并指向下一个链接。

```
class Link
{
public int iData;           //数据
public double dData;        //数据
public Link next;           //指向下一个链接
}
```

这种类的定义有时被称为自引用,因为它包含了一个它自己的类定义。

3. 哈希表

哈希表是一个数据结构,它提供了非常快速的插入和搜索。当第一次听到哈希表,听起来几乎是好得令人难以置信。不管有多少数据项,插入和搜索(有时删除)都可以接近固定的时间:$O(1)$的时间级。实际上,这仅仅是几个机器指令。对于一个哈希表,相对人类用户来说,基本上是瞬时。它是如此之快,计算机程序通常使用哈希表,当它们需要在不到一秒的时间查找成千上万的项目时(如拼写检查)。哈希表比树结构明显快。不仅速度快,哈希表也较易编程。哈希表确实有一些缺点。它们基于数组,而数组一旦创建,扩大就很难。对于某些类型的哈希表,当表变得过满时,性能可能会降低,甚至成为灾难，因此，程序员需要对具体的项目将存储多少数据进行一个相当精确的估算（或为周期性将数据传输到一个更大的哈希表，一个耗时的工序作准备）。

另外，没有便捷的方式可对哈希表实现任何有序的访问（如从最小到最大）。如果需要这个功能，就要查看别的。但是，如果不需要顺序访问各项内容，并且你可以预测你的数据库大小，哈希表的速度和便捷就是无与伦比的。

Unit 9　微软开发者网络

Text A　什么是 MSDN

微软开发者网络（MSDN）是微软的一部分，负责管理协调该公司开发人员和测试人员间的关系：关注操作系统（OS）的硬件开发人员、在不同操作系统平台的开发人员，充分利用 API 和微软许多应用程序的脚本语言的开发人员。这种关系管理分布在不同的场合：网站、简报、开发人员会议、行业媒体、博客和 DVD 发行。关系的生命周期从遗留系统支持到宣传中的潜在产品。

1. 发展史

该服务始于 1992 年，但最初仅在微软开发者网络光盘（CD-ROM）中可用。1993 年增加了二级预订，包括 MAPI、ODBC、TAPI 和 VFW 的软件开发工具（SDK）。2004 年 11 月，MSDN2 作为 Visual Studio 2005 API 信息来源打造，与正在更新网站的代码具有显著的差异，更符合 Web 标准，从而可使用期待已久的支持 API 的改进浏览器替代 IE 浏览器。2008 年，原 MSDN 群退休，MSDN2 成为 msdn.microsoft.com。

1992 年，鲍勃·冈德森在 MSDN 开发者新闻（一个纸质出版物）开始使用假名"Dr. GUI"写专栏。专栏回答 MSDN 订户提出的问题。GUI 博士的漫画是基于 Bob 的照片。当他离开 MSDN 团队后，丹尼斯·克雷恩接手 GUI 博士的角色并加入医疗幽默到栏目中。在他离开后，GUI 博士成为开发者技术工程师原团队（最明显的是保罗·约翰）提供高深的技术文章到数据库的复合代表。总之，这是一个好地方，全方位存放 MSDN 任务。Ken Lassesen 生产的原系统（Panda）发布 MSDN 在互联网上，并以 HTML 格式代替早期的多媒体浏览器引擎。戴尔罗杰森、奈杰尔·汤普森和南希 CLUTs 在 MSDN 团队时出版已发布的 MS 出版社的书籍。截至 2010 年 8 月，微软很少记得那位博士了。

2. 信息服务

微软为软件开发人员提供了一个信息服务。它的主要焦点是微软的.NET 平台，但它也拥有如编程实践和设计模式等领域的文章。网上有许多免费资源，有的则是通过邮件订阅。

根据订阅级别，用户可能会收到微软的操作系统或其他 Microsoft 产品（如 Microsoft Office 应用程序、Visual Studio 等）的早期版本。

大学和高中可以报名参加 MSDN 学术联盟计划，它为他们的计算机科学与工程专业的学生（可能还有其他专业的学生或教师）提供访问一些微软的开发软件。一个 MSDNAA 账户不是一个 MSDN 账户，并且不能用来访问 MSDN 网站或者其订户的下载区域。

3. 软件订阅

MSDN 以往提供订阅包,从而开发人员可以访问和许可使用曾经发布的几乎所有的微软软件。订阅以年论,每年每份花费高达 10 939 美元,因为它是从里到外地提供。订阅持有人(除最低库访问级别)每隔几个星期或数月将通过 DVD 或下载接收到新的 Microsoft 软件。软件一般都在带有特意标明 MSDN 的光盘上,但包含了与零售或批量授权相同的软件,因为它是向公众发布。

虽然在大多数情况下,软件本身的功能与完整产品的完全一样,但在 MSDN 最终用户许可协议禁止在企业生产环境中使用该软件。这是一个法律上的限制,而不是一个技术问题。作为一个例子,MSDN 定期包括最新的 Windows 操作系统(如 Windows Vista 和 Windows 7)、服务器软件(如 SQL Server 2008)、开发工具(如 Visual Studio)、应用程序(如 Microsoft Office 和 MapPoint)。对于需要产品密钥的软件,微软网站根据需求生成。这种包提供了单个计算机爱好者可访问几乎所有微软提供的产品。然而,办公室的 PC 和服务器上运行订阅的 MSDN 软件如果运行在这些机器上没有适当的非 MSDN 许可证,若被抓到,与通过盗版获得的软件,软件许可审计处理结果没有什么不同。

Text B 认识 MSDN

MSDN 的主要网站在 msdn.microsoft.com,它是为开发者社区提供信息、文档、由 Microsoft 和广大社区网民撰写、讨论的网站。最近对论坛、博客、库的注释和社交书签等应用程序的强调和整合正在改变 MSDN 网站的性质,从单向信息服务转变为 Microsoft 和开发人员社区之间的公开对话。在其主站和大部分所列的应用提供 56 种以上的语言内容。

1. System.ComponentModel 命名空间

System.ComponentModel 命名空间提供用于实现组件和控件的运行时和设计时行为的类。本命名空间包括属性和类型转换器实现、数据源绑定和授权组件的基类和接口。

此命名空间的类分为以下 5 类。

- 核心组件类,见组件 Component、IComponent、Container 和 IContainer 类。
- 组件许可,见 License、LicenseManager、LicenseProvider 和 LicenseProviderAttribute 类。
- 属性,见属性类。
- 描述符和持久性,见 TypeDescriptor、EventDescriptor 和 PropertyDescriptor 类。
- 类型转换器,见 TypeConverter 类。

2. AsyncCompletedEventArgs 类

它为 MethodNameCompleted 事件提供数据。

如果你使用的是基于事件的一个同步模式,实现的类将提供 MethodNameCompleted 事件。如果你添加 System.ComponentModel.AsyncCompletedEventHandler 委托处理事件的一个实例,将收到有关同步操作相应事件处理程序方法的参数 AsyncCompletedEventArgs 结果的信息。

客户端应用程序的事件处理程序委托可以检查取消属性,以确定在异步任务是否被取消。

客户端应用程序的事件处理程序委托可以检查 Error 属性,以确定在异步任务的执行过程中异常是否发生。

如果类支持多个异步方法或多次调用同一个异步方法,你可以决定哪些任务通过检查 UserState 属性的值触发 MethodNameCompleted 事件。你的代码将需要跟踪这些令牌(称为任务的 ID),在它们的相应异步任务开始和完成时。

随后的基于事件的异步模式的类可以引发事件,提醒客户端有关悬挂异步操作的状态。如果这个类产生了一个 MethodNameCompleted 事件,可以使用 AsyncCompletedEventArgs 告诉客户有关异步操作的结果。

你可能希望一个异步操作的结果向客户传递比 AsyncCompletedEventArgs 容纳更多的信息。在这种情况下,你可以从 AsyncCompletedEventArgs 类派生自己的类,并提供额外的私有实例变量和相应的只读公共属性。返回属性值前,调用 RaiseExceptionIfNecessary 方法,即使操作被取消或发生错误。

3. IComponent 接口

IComponent 接口提供了所有组件所需的功能。

Component 是 IComponent 的默认实现,并作为公共语言运行库的所有组件的基类。

可以在容器中包含部件。在这种情况下,容纳指逻辑容纳,而不是视觉容纳。你可以在各种情景下使用组件和容器,包括可视的和非可视的。

System. Windows. Forms. Control 从 Component 组件继承下来,是 IComponent 的默认实现。

一个组件与它的容器相互作用主要通过一个容器提供的 ISite,这是特定容器的每个组件信息构成的信息库。

要成为一个组件,类必须实现 IComponent 接口,并提供一个无需参数或单个参数 IContainer 类型的基本构造函数。有关 IComponent 实现的更多信息,请参阅使用组件编程。

Unit 10　编　译　原　理

Text A　代码优化的科学

术语"优化"在编译器设计中指的是一个编译器试图产生比直接代码更高效的代码。因此,"优化"一词用词不当,因为没有办法保证编译器产生的代码与任何其他代码执行相同的任务一样快或更快。

现在,代码优化的编译器执行已经变得更重要、更复杂了。更复杂是因为处理器架构变得越来越复杂,产生更多的机会改善代码执行方式。更重要是因为大规模并行的计算机需要大量的优化,或它们的性能取决于执行顺序。随着多核机器(计算机芯片上有多个处理器)的出现,所有的编译器将不得不面对充分利用多处理器机器优势的问题。

这是很难,不是不可能的,但要建立一个强大的编译器"黑客"。因此,围绕优化代码的问题,扩展的和有用的理论已被建立起来。采用严格的数学的基础,让我们表明,优化是正确的,对所有可能的输入它产生了理想的效果。我们将看到,从第9章,如果编译器要生成优化好的代码,那么图、矩阵和线性程序等模型是多么重要。

另一方面,仅靠纯理论是不够的。就像许多现实世界的问题,没有完美的答案。事实上,大多数的我们在编译器优化提出的问题是不可判定的。其中,在编译器设计中一个最重要的技能是确定要解决真正问题的能力。我们从对程序的行为有很好的理解入手,通过实验和评估验证我们的直觉。编译器优化必须满足以下设计目标。

优化必须正确,也就是说,保持已编译的程序的含义。

优化必须改善许多程序的性能。

编译时间必须合理。

所需的工程工作必须是可控的。

无论如何强调正确性,都不过分。编译器产生快速的代码,如果生成的代码不一定正确,就是没有意义的!优化的编译器如此难以保证正确,我们敢说,没有优化的编译器是完全没有错误的!因此,编写编译器最重要的目标是,它是正确的。

第二个目标是,编译器必须有效地改善多输入程序的性能。通常表现为程序执行的速度。尤其是在嵌入式应用中,我们不妨尽量减少生成代码的大小。而在移动设备情况下,也期望是,代码减少功耗。通常,同一优化既加快了执行时间,也降低了功耗。除了性能,有关错误报告和调试的方面,可用性也很重要。

第三,我们需要保证编译时间短,支持快速开发和调试周期。因机器变得更快,此要求已变得更容易满足。通常,一个程序首先必须经过程序优化进行开发和调试。不仅是编译时间减少了,更重要的是,未优化方案比较容易调试,因为由编译器引入的优化经常混淆源代码和目标代码之间的关系。打开的编译器优化选项有时暴露源程序中的新问题;因此,必须再次对优化的代码进行测试。需要额外的测试有时阻止在应用中使用优化,尤其是如果它们的性能并不重要时。

最后,一个编译器是一个复杂的系统;我们必须保持系统简单,以确保编译器的工程和维护成本是可控的。还有,我们可以实现程序的优化无限化,它需要付出不平凡的努力创建一个正确的和有效的优化。我们必须优先考虑的优化,仅是针对在实践中遇到的源程序可取得最大效益的工作。

因此,在研究编译器时,我们不仅学习如何构建一个编译器,同时也学习解决复杂和开放性问题的一般方法。编译器开发使用的方法既包括理论,又包括实验。我们通常基于直觉认可的什么是重要问题开始进行规划。

Text B　计算机结构优化

计算机体系结构的快速发展导致新的编译器技术无限的需求。几乎所有的高性能系统都采取同样的两种基本技术优势:并行性和存储器层次结构。并行性可以在多个级别中找到:在指令级,其中多个操作同时执行;在处理器级,其中同一应用程序的不同线程在不同的处理器上运行。内存层次结构的基本限制,我们可以建立非常快的存储或量非常大的存储,而不是存储既快速,量也大的响应。

所有的现代微处理器都开发指令级并行性。然而,这种并行性可以向程序员隐藏。

编写程序时,总认为所有指令是串行执行的;硬件动态检查顺序指令流中的依赖关系,尽可能发布它们并行执行。在一些情况下,机器包括硬件调度器,可以改变它的指令排序,也可以增加并行性。无论是硬件重新排序,或没有,编译器都可以对指令重新排序,使指令级并行更有效。

指令级并行也可以明确地出现在指令集上。VLIW(超长指令字)机器具有可以并行发出多个操作的指令。英特尔 IA64 是这种架构的著名例子。所有的高性能通用微处理器也包括可以在同一时间上操作数据矢量的指令。这样的编译器技术已被开发,从顺序程序自动地产生代码用于这样的机器。

多处理器已很普遍。即使个人计算机,也通常具有多个处理器。程序员可以为多处理器编写多线程代码,或者通过从传统的顺序程序编译器自动生成并行代码。这样的编译器对程序员隐藏了许多细节,在程序发现并行性,横跨不同机型的分发计算,并尽量减少处理器之间的同步和通信。许多科学计算和工程应用都是计算密集的,可以从并行处理大大受益。并行化技术已经发展到自动将串行科学程序翻译成多处理器的代码。

存储器层次结构包含存储速度和容量大小不同水平的几个层次,带有最接近处理器而速度最快的,但容量最小的层次。平均存储的存取时间将减少,如果大部分的访问是由层次结构速度更快的存储器满足程序需要。并行性和存储器层次结构的存在都可提高机器的潜在性能,但它们必须有效地通过编译器实现应用程序的实时性能。

所有计算机都有存储器层次结构。处理器通常含有小数目的寄存器、几百上千字节、不同水平的几千到兆字节高速缓存、兆字节到千兆字节物理存储器,以及从吉字节甚至更多的最终辅助存储。相应地,层次结构的相邻水平之间的访问速度不同幅度可以达到两个或 3 个数量级。一个系统的性能经常不受限于处理器速度,而是取决于存储器子系统的性能。传统编译器重点优化处理器的执行,现在更强调放置在哪,使得存储器层级更有效。

有效使用寄存器在优化方案中可能是一个最重要的问题。不像寄存器可以在软件进行显式管理,高速缓存和物理存储器隐藏在指令集并由硬件管理。现已发现,在一些情况下,通过硬件实现的高速缓存管理策略并不有效,尤其是在科学代码具有大的数据结构(典型的矩阵)时。有可能通过改变数据的布局,或改变存取数据的指令顺序,以提高存储器层次结构的有效性。我们还可以改变代码的布局,以提高指令缓存的有效性。

Unit 11　操作系统

Text A　操作系统概述

1. 什么是操作系统

操作系统是软件,由计算机上运行的程序和数据组成,主要管理计算机硬件并为各种应用软件有效执行提供公共服务。

管理硬件的功能如输入、输出和存储器分配,操作系统充当应用程序和计算机硬件之间的中介,虽然应用程序代码通常由硬件直接执行,但会频繁调用操作系统或被它中断。几乎在任何包含一台计算机的设备上,都可以找到操作系统,从蜂窝手机和视频游戏机到超级计算机和 Web 服务器。

2. 操作系统的历史

20 世纪 50 年代初,一台计算机同一时间只能执行一个程序。每个用户单独使用计算机,并会在预定的时间通过穿孔纸卡片和磁带提供程序和数据。程序将加载到该设备,机器将一直设置为工作状态,直到程序完成或崩溃。程序一般可以调试,通过使用前面控制板上的切换开关和面板灯。据说,阿兰·图灵是早期的曼彻斯特马克 1 号机上的主人,他已经从通用图灵机的原理得出操作系统的原始概念。

以后的机器附带软件库,这将链接到用户的程序协助一些操作,如输入和输出,将人可读的符号代码转换成计算机代码。这是现代操作系统的起源。但是,计算机在同一时间仍然运行单个作业。在英国剑桥大学,在同一时间作业队列的磁带上挂着不同颜色的衣服夹子,表明作业优先级。

3. 大型机

经过 20 世纪 50 年代,在操作系统领域开创了许多主要功能,包括批量处理、输入输出中断、缓冲、多任务处理、后台、运行时库、链接加载,以及在文件中记录排序程序。这些特征可作为应用程序员的选项,应用软件可纳入或不纳入,而不是所有应用都使用一个独立的操作系统。1959 年,作为 IBM 704、709 和 7090 主机的综合工具的操作系统发行了。

20 世纪 60 年代,IBM 的 OS/360 推出了一个操作系统的概念,涵盖了整个产品线,对 System/360 取得成功至关重要。IBM 目前的大型机操作系统都是这种原始系统的远亲后代,并且 OS/360 上的应用程序仍可以在现代机器上运行。20 世纪 70 年代中期,MVS、OS/360 的后裔第一个实现了使用 RAM 作为数据的透明缓存。

4. 微电脑

第一代微型计算机还没有精心制作操作系统的能力或需求,如没有开发用于大型机和迷你机系统,而是开发简约操作系统,往往从 ROM 加载并称为监视器。一个著名的、早期的、基于磁盘的操作系统是 CP/M,它得到许多早期的微型计算机支持,并被 MS-DOS 高度模仿,变得广受欢迎,作为 IBM PC 操作系统的选择(IBM 版称为 IBM DOS 或 PC DOS),是微软的成功产品。20 世纪 80 年代,苹果电脑公司(现在的苹果公司)放弃了其广受欢迎的 Apple II 系列单片机,创新地将图形用户界面(GUI)Mac OS 引入到苹果 Macintosh 计算机。

5. 操作系统的实例

(1) 微软 Windows

微软 Windows 是在个人计算机上最常用的特有操作系统家族。它是个人计算机中最常见操作系统的系列,目前市场占有率约 90%。Windows 系列中使用的最广泛的版本是 Windows XP,其于 2001 年 10 月 25 日发布。最新的版本是适用于个人计算机的 Windows 7 和适用于服务器的 Windows Server 2008 R2。

(2) UNIX 和 UNIX 类操作系统

肯·汤普逊写 B 代码,主要是基于 BCPL,基于他在 MULTICS 项目经验,他使用它编写 UNIX。用 C 语言代替 B 语言,UNIX 发展为大型、复杂的相互关联的操作系统家族,已影响到每一个现代操作系统。UNIX 类家族是一组不同的操作系统,有几个主要的子类,包括 System V、BSD 和 GNU/Linux 操作系统。名称"UNIX"是 The Open Group 的商标,用于与符合其定义的任何操作系统使用的许可证。"UNIX 类"通常指类似于原 UNIX 操作系统组成的大型集合。

(3) Mac OS X

Mac OS X 是由苹果公司开发、营销和销售的专有图形操作系统系列产品,最新的版本是所有目前推出的 Macintosh 计算机预先安装的。Mac OS X 的前身是原来的 Mac OS,自 1984 年以来,它是苹果机的主流操作系统。Mac OS X 是建立在 UNIX 操作系统开发的技术,这由 Next 公司开发始于 20 世纪 80 年代后期,直到 1997 年初苹果公司购买了该公司。

1999 年首次发布 Mac OS X 服务器 1.0 操作系统,2001 年 3 月发布了面向桌面版本(Mac OS X 10.0 版),此后,6 个不同的 Mac OS X "客户"和"服务器"版本已经发布,最近一次是 Mac OS X v10.6,于 2009 年 8 月 28 日首次提供。Mac OS X 版本以大型猫科动物的名字命名;Mac OS X 的最新版本为"雪豹"。

(4) 其他

目前仍在 Niche 市场使用的旧操作系统包括来自 IBM 和微软的 OS/2、Mac OS、非 UNIX 基础的苹果 Mac OS X、BeOS、XTS-300。有些操作系统(如著名的 Haiku、RISC OS、MorphOS 和 AmigaOS 4)继续发展为发烧友社区和专业应用的少数人平台。从 DEC 形成的开放 VMS,仍由惠普积极发展。然而,还有其他操作系统几乎只用在学术界,作为操作系统教育或在其上做操作系统的概念研究。同时履行这两大功能的典型例子是 MINIX,而如 Singularity 纯粹用于研究。

Text B BIOS 或 CMOS 设置

BIOS 是基本输入输出系统的缩写,是内置的软件,决定计算机可以做什么(如不从磁盘访问程序)。在 PC 上,BIOS 包含所有控制键盘、显示屏、硬盘驱动器、串行通信,以及一些辅助功能所需的代码。所有选项如图 11-1~图 11-3 所示。

BIOS 通常放置在计算机自带的(这通常称为 ROM BIOS)ROM 芯片中。这确保了 BIOS 将始终可用,并且不会被磁盘故障损坏。这也使得计算机自身引导有可能。因为 RAM 比 ROM 的速度快,虽然很多计算机厂商设计系统,每次计算机启动时,将 BIOS 从 ROM 复制到 RAM 中,这称为投影。

由于存在各种各样的计算机和 BIOS 厂商,所以有很多方法可以进入 BIOS 或 CMOS 设置。下面是这些主要方法的一个列表,以及用于输入 BIOS 设置的其他建议。

值得庆幸的是,过去几年制造的计算机在引导过程中已经可以让你按以下 5 个键之一进入 CMOS。

F1、F2、Del、Esc、F10

通常用前 3 个键之一。

用户在计算机启动时看到类似于下面示例的消息时,会知道何时按下此键。某些较旧的计算机还可能显示一个闪烁块,以指示何时按 F1 或 F2 键。

```
Press <F2> to enter BIOS setup (按<F2>键进入 BIOS 设置)
```

成功进入 CMOS 设置后,会看到类似于图 11-1 的屏幕。

参 考 文 献

[1]　Cohen, Bernard, Howard Aiken. Portrait of a computer pioneer[M]. Cambridge, Massachusetts: The MIT Press, 2000.

[2]　Stokes, Jon. Inside the Machine: An Illustrated Introduction to Microprocessors and Computer Architecture[M]. San Francisco: No Starch Press, 2007.

[3]　Alfred V Aho, Monica S Lam, Ravi Sethi, et al. Compilers: Principles, Techniques and Tools[M]. New Jersey: Addison-Wesley, 2006.

[4]　Abraham Silberschatz, Peter B Galvin, Greg Gagne. Operating System Concepts[M]. 9th ed. Toronto: John Wiley & Sons, Inc, 2012.

[5]　Christopher Negus. Linux Bible[M]. 2010 ed. Toronto: Wiley, 2010.

[6]　Andrew S Tanenbaum. Modern Operating Systems[M]. 4th ed. New Jersey: Prentice Hall. 2014.

[7]　Ryan Asleson, Nathaniel T Schutta. Foundations of Ajax[M]. New York: Springer-Verlag New York Inc., 2005.

[8]　Sommerville. Software Engineering[M]. 8th ed. New Jersey: Addison Wesley, 2006.

[9]　Grady Booch. The Unified Modeling Language User Guide[M]. 2nd ed. New Jersey: Addison-Wesley, 2008.

[10]　Jon Kleinberg. Algorithm Design[M]. Pearson: Addison Wesley, 2005.

[11]　Mitchell Waite. Data Structures and Algorithms in Java[M]. Toronto: Wiley, 2010.

[12]　Paul DuBois. MySQL Cookbook[M]. 2nd ed. Sebastopol: OReilly Media, 2007.

[13]　Micbael Barr. Programming Embedded Systems in C and C++[M]. Sebastopol: O'Reilly Media, 1999.

[14]　Daniel W Lewis. Fundamentals of Embedded Software: Where C and Assembly Language Meet [M]. New Jersey: Prentice Hall, 2012.

[15]　James Keogh. J2ME: The Complete Reference[M]. New York: McGraw-Hill, 2003.

图书资源支持

感谢您一直以来对清华版图书的支持和爱护。为了配合本书的使用,本书提供配套的资源,有需求的读者请扫描下方的"书圈"微信公众号二维码,在图书专区下载,也可以拨打电话或发送电子邮件咨询。

如果您在使用本书的过程中遇到了什么问题,或者有相关图书出版计划,也请您发邮件告诉我们,以便我们更好地为您服务。

我们的联系方式:

地　　址:北京市海淀区双清路学研大厦 A 座 714

邮　　编:100084

电　　话:010-83470236　010-83470237

客服邮箱:2301891038@qq.com

QQ:2301891038(请写明您的单位和姓名)

资源下载: 关注公众号"书圈"下载配套资源。

资源下载、样书申请

书圈

图书案例

清华计算机学堂

观看课程直播